Communicating Science Effectively

A practical handbook for integrating visual elements

J.E. Thomas, T.A. Saxby, A.B. Jones, T.J.B. Carruthers, E.G. Abal, W.C. Dennison

www.ian.umces.edu   www.umces.edu

University of Maryland
CENTER FOR ENVIRONMENTAL SCIENCE

www.loicz.org   www.internationalriverfoundation.com.au

International
Riverfoundation

IWA
Publishing

Published by IWA Publishing, Alliance House, 12 Caxton Street, London SW1H 0QS, UK
Telephone: +44 (0) 20 7654 5500; Fax: +44 (0) 20 7654 5555; Email: publications@iwap.co.uk
Web: www.iwapublishing.com

First published 2006
© 2006 IWA Publishing

Printed by Alden Press, Oxford, UK

Disclaimer

British Library Cataloguing in Publication Data
A CIP catalogue record for this book is available from the British Library

Library of Congress Cataloging- in-Publication Data
A catalog record for this book is available from the Library of Congress

ISBN: 1843391252
ISBN13: 9781843391258

Table of contents

Preface

This handbook is designed by scientists, for scientists. It is our thesis that good science needs to be accompanied by good communication in order that the science can contribute to furthering our understanding of nature and to solving societal problems, such as environmental degradation. Most scientific training is focused on the technical and analytical skills needed to obtain and interpret scientific data. Conversely, little training is devoted to communicating science, and the little training that occurs is focused on communicating exclusively to peer scientists. Thus, this handbook and accompanying web material are designed to focus on the communication of scientific findings to both peer and non-peer scientists and to the wider audience of resource managers and interested lay people. The term 'science communication' often refers to journalists who write or produce science stories for mass media consumption. Our efforts are focused on practicing scientists who need to improve their skills in communication, but perhaps journalists will also find some value in the topics presented here.

We have come upon the various principles and examples used in this handbook through our own experiences in attempting to communicate science. Every effort is made to analyze the successful aspects of science communication that we have been evolving, in order that we can better give the reader the guidance and, hopefully, the inspiration to become a more effective science communicator. Thus, this is a practical and applied approach to the subject, not a theoretical or particularly academic treatise. This is a work in progress and we invite feedback and critique. We hope that you find this a useful handbook and if it results in an improvement in your ability to communicate science, then we have accomplished our intended goal.

William C. Dennison, Jane Thomas, Tim Carruthers, Tracey Saxby, Adrian Jones, Eva Abal

Cambridge, Maryland
May 2006

The alphabet is a funnel …
John Culkin

… but visual language unleashes the full power of communication.
Robert E. Horn

1.

Why is effective science communication important?

Scientific discovery can result in large lifestyle and philosophical changes in society. Throughout most of history, science has provided new opportunities, such as discovery of new lands, new resources, new technologies, or new insights (e.g., the formation of the planets). For much of that history, science has been carried out by the intellectual and social elite, so dissemination of new ideas was relatively easy amongst this small sector of society. However, in the 21st century, there are some fundamental differences to the old model of scientific discovery and information dissemination. There are now many more scientists working in disparate fields, and most of these people are very specialized—no longer are writers-philosophers and scientists in the same community. There are now accepted and formal ways of communicating within the scientific community, mainly through publishing articles in journals and giving presentations to colleagues at large meetings.

Throughout the 20th century, science was proclaimed as the solution to the problems of land degradation and pollution resulting from earlier discoveries during the agricultural and industrial revolutions. As a result, there is now an increasing focus on funding for science being linked to providing practical solutions to environmental problems. This creates a dilemma, for while excellent science can be conducted, science alone will not create widespread change, mainly because the channels to use this information and create change are poorly developed. In order to create changes in behavior and beliefs of the general public, broader and more effective communication of the new scientific insights being gained is required. Even where the solutions to environmental problems are clear, management, political, and ultimately public support are needed to implement the (usually) expensive solutions. Therefore, utilizing our current research effectively will require new tools to facilitate effective communication, not only to scientific peers, but also to managers, government, and ultimately, the general public.

The essence of effective science communication is the development of *content-rich, jargon-free, communication-based* materials. *Content-rich* refers to communication that is replete with data and ideas. *Jargon-free* refers to the elimination of shorthand notation that scientists use to communicate within their peer groups—this means removing acronyms and maintaining a common language basis for explaining concepts. *Communication-based* refers to focusing on the intended audience and providing an even broader base of accessibility for a wider audience.

Creating social change and solving environmental problems requires knowledge and power. Scientists have high knowledge but little power, whereas politicians have a lot of power but often little knowledge of many environmental problems. Effective science communication can facilitate this link between knowledge and power, informing and empowering the public to produce social change.

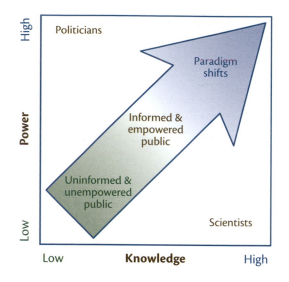

EFFECTIVE SCIENCE COMMUNICATION CHANGES SOCIETAL PARADIGMS

Science has progressed over time with a series of paradigm shifts. These shifts occur when scientific understanding is effectively communicated to society. In an attempt to predict the next major shift, an analysis of the history of scientific paradigms was conducted. In the words of Winston Churchill, "The farther backward you can look, the farther forward you are likely to see." Over the past 500 years, a series of major paradigm shifts have occurred. Dividing the historical time-line into 50-year periods, 10 paradigm shifts have occurred since the year 1500. The first of these (1500–1550) came in astronomy from the work of Nicolas Copernicus, who postulated that the earth was not at the center of the solar system—rather, that the earth revolved around the sun. This was supported by the observations and writings of Galileo Galilei (1550–1600), who believed that the heavenly bodies consisted of physical matter rather than ethereal substances. Both Copernicus and Galileo were responding to the impetus of a need to understand where the earth was placed in the broader spectrum of the universe.

The next major paradigm shift (1600–1650) was also in astronomy. Johann Kepler formulated three laws of planetary motion, now known as Kepler's laws, which stated that planets moved in elliptical orbits, not circular orbits. Isaac Newton precipitated another paradigm shift (1650–1700) with his book on the principles of mathematics, in which he demonstrated that there were universal physical laws (e.g., gravity), which supplanted the belief that the forces of nature were only affected through physical contact. In the period 1700–1750, as more of the earth was explored, the diversity of life became evident and a need to categorize living things was evident. As a response, Carl Linnaeus and his students developed a uniform method of naming organisms, still in use today, replacing the multiple names for the same organism that previously existed.

In the period 1750–1800, the French nobleman and chemist Antoine Lavoisier disproved the phlogiston theory of combustion. This earlier theory stated that all flammable materials contain *phlogiston*, a substance without color, odor, taste, or weight that is released during burning. Instead, Lavoisier showed that combustion requires oxygen, setting the stage for a new theory of what happens when objects burn, and identifying and naming oxygen in the process.

Yet another major paradigm shift occurred during 1800–1850 with respect to the earth's formation. Charles Lyell postulated that the earth was shaped by gradual processes, or uniformitarism, rather than by catastrophic events. This followed Hutton's theory that the age of the earth was much greater than the accepted 6,000 years.

The evolution period (1850–1900) revolutionized the way people thought about the origin of the human species. Charles Darwin's books on evolution were best-sellers and sparked considerable debate throughout society. A key aspect of Darwin's contribution was his ability to communicate the ideas of natural selection and evolution to society through his writings.

The physics period (1900–1950) was the era of substantial discoveries in the nature of matter. Albert Einstein's theory of relativity provided a paradigm shift in the view of matter and energy, postulating that matter and energy were interchangeable. This improved understanding of matter provided the basis for nuclear physics and eventually led to atomic power and atomic bombs. The biology period (1950–2000) was stimulated by the elucidation of the structure of DNA (deoxyribonucleic acid) by James Watson and Francis Crick. The ensuing advances in molecular biology led to biotechnology, the human genome project, and new insights into the evolutionary relationships of living things.

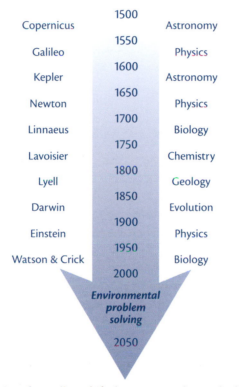

A series of paradigm shifts have occurred over the last 500 years. Communicating science effectively results in large lifestyle and philosophical changes in society.

The last two centuries have seen the splitting of philosophy and literature from science. Currently, while scientists have high science quality, they generally have low communication quality, while modern philosophers and orators have high communication quality and generally low science quality. Effective science communication is required to bridge the gap between modern scientists and modern philosophers and orators to facilitate new paradigm shifts in societal thinking.

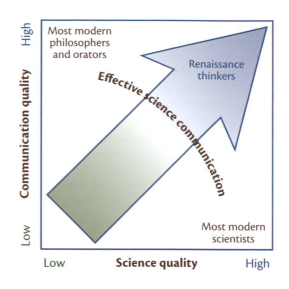

EFFECTIVE SCIENCE COMMUNICATION CAN MAKE YOU A BETTER SCIENTIST

Effective communication is an important part of doing science. Many great scientists are also great communicators. Consideration of the communication aspect of scientific research can lead you to producing better and more complete results. There are several ways in which attention to the communication aspects of science can improve your science. *Completeness*—envisioning the story that is being conveyed can lead to a more comprehensive research program in which each element of the story is addressed. Having conceptualized the storyline for the science communication product, you can fill the holes or gaps in order to make the story complete. *Context* is identifying linkages and developing comparisons that can lead to important insights. The search for explanations of temporal or spatial comparisons often leads to a fresh perspective on the data. *Visualization* is a powerful communication tool that can provide unique insights. For example, production of a map using overlays of different elements can provide linkages that may not otherwise be obvious. An aerial photograph combined with a conceptual diagram provides another example of visual elements that would not be as powerful individually as they are when combined. *Synthesis* is achieved by combining different data or approaches, which can lead to novel insights. For example, combining all the data on a topic and developing either a mathematical depiction or a correlation with another feature can create important benchmarks, such as the Redfield Ratio of elemental carbon:nitrogen:phosphorus concentrations (106:16:1) in which global values for both phytoplankton and water column concentrations were calculated. This ratio has been used extensively by oceanographers, but also modified and applied to other organisms and ecosystems.

THE ESSENCE OF SCIENCE COMMUNICATION

Much of this handbook is concerned with the art of science communication. In addition to the art, you will need to find the essence of science communication. The essence includes allowing adequate time to produce science communication products and also giving yourself enough quality time in which your creative efforts can be made and integrated into the science communication products. Balancing time demands is always an issue, but the use of quality time to develop good science communication products will be appreciated by your audience. This quality time needs to involve creating opportunities for critical feedback, so that a small group of friends or colleagues can help improve the draft stages of your project. Feedback and revision is essential, and feedback often needs to be actively sought. In the planning process, time allotted to feedback and associated recommended revisions needs to be built in, otherwise the project may drift along in the absence of a fixed timeline. The concept of multiple revisions of science communication products is not often realized. However, traditional scientific writing has both an informal and a very formal review process, and it is expected that a published paper will have undergone a thorough evaluation and multiple revisions. This review of science communication can be made both internally (e.g., lab groups or graduate student seminar series) or externally (e.g., editors, reviewers).

It is important to allow your enthusiasm for your subject to come forward. Expressing enthusiasm is often discouraged on the grounds that it is a distraction to the dispassionate and objective scientific analysis. However, enthusiasm about the topic should not influence your scientific approach—rather, sharing the excitement of discovery and enthusiasm about learning serves to contagiously entrain the audience. If the audience senses that the scientist presenting findings does not really care about the topic, then they will invariably wonder why they should care. If your enthusiasm is apparent, your audience will be more likely to attempt to comprehend your message.

Good science communication requires attention to both the science and the presentation. If the science is not good, it does not matter how well you dress it up—it is still not good science. If the science is good, but it is not presented well, it loses its power and impact. In the worst case scenario, this becomes an indulgent hobby for the self-edification of the scientist and is not used to build the body of knowledge. The goal is to end up with good science that is effectively communicated. In general practice, the vast majority of scientific effort is in the collection and analysis of data, with little time or resources devoted to the communication of science. Rather than science communication being an afterthought, it is essential to factor in the time and resources that are needed for developing a quality communication product.

FURTHER INFORMATION

Abal EG, Bunn SE, & Dennison WC (eds). 2005. *Healthy Waterways, Healthy Catchments: Making the connection in South East Queensland, Australia.* Moreton Bay Waterways and Catchments Partnership, Brisbane, Queensland, Australia.

Cribb J, & Hartomo TS. 2002. *Sharing knowledge: a guide to effective science communication.* CSIRO Publishing. Collingwood, Victoria, Australia.

Dennison WC, & Abal EG. 1999. *Moreton Bay Study: a scientific basis for the Healthy Waterways campaign.* South East Queensland Regional Water Quality Management Strategy, Brisbane, Queensland, Australia.

Montgomery SL. 2003. *The Chicago guide to communicating science.* The University of Chicago Press, Chicago, Illinois, U.S.A.

South East Queensland Regional Water Quality Management Strategy Team. 2001. *Discover the waterways of south-east Queensland.* South East Queensland Regional Water Quality Management Strategy, Brisbane, Queensland, Australia.

Tufte ER. 1990. *Envisioning information.* Graphics Press, Cheshire, Connecticut, U.S.A.

Tufte ER. 1997. *Visual explanations: images and quantities, evidence and narrative.* Graphics Press, Cheshire, Connecticut, U.S.A.

Tufte ER. 2001. *The visual display of quantitative information.* Graphics Press, Cheshire, Connecticut, U.S.A.

Valiela I. 2001. *Doing science: design, analysis, and communication of scientific research.* Oxford University Press, New York, New York, U.S.A.

… from all the facts assembled, there arises a certain grandeur.

Aristotle

2.

What is effective science communication?

Effective science communication is the successful dissemination of knowledge to a wide range of audiences, from specialist scientists, through managers and politicians, to the general public. Many scientists believe that doing excellent science is enough and that this knowledge will be found and used at the appropriate time. Unfortunately, the public, politicians, and even environmental managers rarely read journal articles or highly specialized books, so these media alone do not constitute effective science communication. Increasingly, scientists are called upon to comment on current environmental problems and search for solutions—however, they are often left lacking tools to communicate the knowledge that they have, especially in the face of the uncertainty inherent in the scientific process. It is the nature of science that a scientist can never be 100% certain, which is problematic to those responsible for decision-making. However, with appropriate communication tools it is possible for scientists to explain their messages to a broader audience, creating greater understanding and demystifying both scientific knowledge and the scientific process. Only when effective science communication is achieved will the relevance of science to society in general increase.

PROVIDING SYNTHESIS, VISUALIZATION, AND CONTEXT

The key elements of science communication are synthesis, visualization, and context. Raw data do not provide much insight to anyone except perhaps the investigator collecting the data. Rather, data that have been analyzed, interpreted, and synthesized are needed for meaningful science communication products. Visualization is key, as the audience must be able to see the *who, what, where, when,* and *how* of the data that are used to support the ideas, so that the scientist can then tell them *why*. A common strategy is to not provide the data, and inform the audience that the data are very complicated and that the listeners should just trust the scientist that the ideas are supported by the missing data. This is a flawed strategy and often means that the scientist has not worked hard enough to develop effective communication devices. Making a point with the data visualized is very powerful. The audience needs to be able to see and interpret the data themselves—they can be guided through the data, but they need to know that the data exist. Context provides answers to the important questions 'Why should we care?', or more simply, 'So what?' Relate to the audience, giving them the big picture while being locally relevant. Context includes using comparative data so that specific examples can be characterized as 'high' or 'low' relative to regional or global extremes. Context also lets people understand why you are measuring what you are measuring, or why you care about a certain issue.

SIMPLIFYING YOUR TERMS BUT NOT YOUR CONTENT

Do not dumb it down—instead, raise the bar. You can assume your audience has very little prior knowledge of your particular study, but also assume that the people in your audience are intelligent, knowledgeable people who have the capacity to be informed. If your audience is naïve, that does not mean they are stupid. Effective science communication should aim to educate the audience and bring them up to your level. People can generally grasp even complex concepts if they are communicated effectively.

ASSEMBLING SELF-CONTAINED VISUAL ELEMENTS

Science communication relies on the use of images, maps, photos, tables, figures, video clips, and conceptual diagrams. Science communication principles can be applied to all science communication products, including proposals and papers published in peer-reviewed scientific journals, newsletters, books, videos, mass media, and effective communication at meetings and conferences. Start a personal resource library containing different visual elements that will help you to communicate your research. Creating effective graphics and illustrations is one of the most time-consuming tasks, but appropriate use of illustrations will dramatically improve the communication of your story—a picture is worth 1,000 words!

Conceptual diagrams

Conceptual diagrams help to clarify thinking—words can be ambiguous but an image commits to the message being portrayed. Use of symbols in conceptual diagrams is an ancient communication technique to depict unequivocal messages. Conceptual diagrams facilitate communication, both one-way (the presentation of the idea), and two-way (idea development). They can be used to identify gaps in, priorities of, and essential elements of knowledge by communicating concepts, summarizing information, and indicating key processes. Conceptual diagrams provide synthesis, visualization, and context. For more information on conceptual diagrams, see Chapter 4.

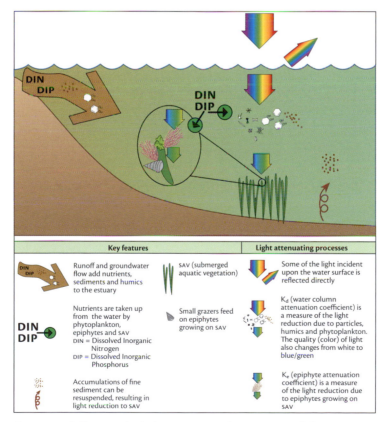

Conceptual diagram depicting processes that attenuate light.

Satellite photos and maps

Satellite photos and maps provide geographic context and are information-rich. Depending on scale, satellite photos provide extra information by showing topography and land use, as well as water clarity, depth, and movement. A series of satellite photos or maps can be useful in accentuating differences and tracking temporal changes.

Maps are very useful when communicating science. They provide information about experimental or sampling locations, as well as the local geographical context of the area.

Photographs

Photos provide unique information. Aerial photos can serve as excellent site-scale maps. Experimental photos can depict methods and display visible impacts of experimental manipulation, especially when taken at different times during a study. Photos can also provide a context for research—for example, scientists who do not work in the aquatic science field may be naïve about underwater habitats. The use of historical photos can help verify conjecture and anecdotal evidence. Such photos are also useful in involving the community by making them aware of changes in their environment. Photos can also help to determine targets for restoration efforts.

Study site on Amity Banks, eastern Moreton Bay.

Cyanobacterium *Lyngbya majuscula* growing on the seagrass *Zostera capricorni*.

CHRIS ROELFSEMA

An example of using photographs to portray information. The aerial photo on the left provides the larger-scale context by showing the location of the study site, and even the cyanobacterial bloom is visible as the darker-colored areas. The photo on the right contains the detail of the cyanobacterial bloom.

Video clips

Short video clips inserted into your presentations can help to tell your story by capturing motion, perspective, and sound. They can be especially effective for people who have not seen the habitat or organism which you are studying.

Your data

Your data are the backbone of your research, and can be displayed in a variety of ways. Graphs and tables are the most common forms of data presentation.

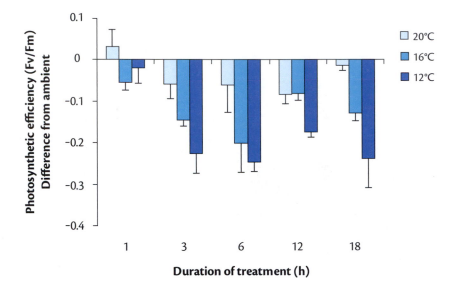

Graphs are the most common way of displaying data, and when formatted correctly, are a very effective way of communicating your message.[1]

Reporting region	Water Quality Index	Chlorophyll a	δ^{15} Nitrogen	Dissolved oxygen	Total nitrogen	Total phosphorus	Secchi depth
Assawoman Bay	0.56	1	0.63	1	0.45	0.06	0.15
Chincoteague Bay	0.42	0.40	0.76	0.87	0.40	0	0.05
Chincoteague Appendix	0.62	0.86	0.80	1	1	0	0.02
Isle of Wight Bay	0.69	1	0	1	1	0.44	0.70
Newport River	0.36	0	0.83	1	0.21	0.07	0
Newport Bay	0.33	0.11	1	0.89	0	0	0
Sinepuxent Bay	0.68	0.88	1	0.96	0.86	0.21	0.22
St. Martin River	0.29	0.45	0.25	1	0.03	0	0

Tables are an excellent way of presenting a lot of data, especially when you are making comparisons between many variables.[2]

Visual elements can be combined to provide unique information. For example, a combination of a photo and a conceptual diagram can effectively orient the audience to your study site, or explain methodology. Photos and graphs together can help with the visualization of your results. Results can be overlaid on to maps, which helps the audience envisage the overall context of your results.

On the front page, the satellite photo gives the big-picture context of Hurricane Isabel over the east coast of the U.S.A. The graph of the storm surge and the photos of flooding tell a more regional story.

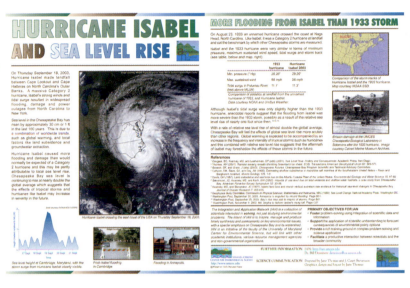

On the back page, a comparison between Hurricane Isabel and a 1933 hurricane is given by a table of data comparing the two storms, a map showing the tracks of the hurricanes, and an historical photo showing erosion damage after the 1933 storm.

In the center spread, the issue of sea level rise is addressed by graphs showing projected temperature and sea level rise, as well as long-term data from a tide gauge in Baltimore. A conceptual diagram with a self-contained legend depicts the processes contributing to sea level rise.

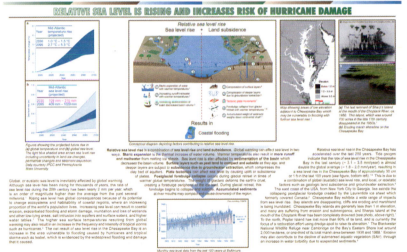

The local context to Chesapeake Bay is given here by a map showing the elevation of the coastal plain surrounding the bay. An historical photo and a recent photo graphically depict the results of sea level rise and erosion in Chesapeake Bay.

In this science newsletter, all the visual elements discussed above—conceptual diagrams, satellite photos, maps, photos, graphs, and tables—along with a judicious amount of text, are used to communicate the story about sea level rise and the flooding associated with Hurricane Isabel in September, 2003.[3]

ELIMINATING JARGON, DEFINING YOUR TERMS, AND MINIMIZING ACRONYM USAGE

Science is full of jargon, which Webster's Dictionary defines as *nonsensical, incoherent, or meaningless talk* and also as *the specialized or technical language of a trade, profession, or similar group*. These two definitions immediately illustrate why jargon can confuse or alienate members of your audience who are not familiar with your field of study. Instead of using jargon, translate it into terms that can be understood by somebody who has no background in your field.

Do not assume that your audience is familiar with all the terms that you use when describing your research. Define all your terms so that they are understandable by people who have no background in your research area. Do not shy away from presenting complex ideas—however, the details of methods and techniques may not be relevant. Often in science, the concepts are rather simple but the circuitous manner of collecting relevant data using technologies that change rapidly is not simple.

An acronym is *a word formed from the initial letters of a name, such as* **WAC** *for* **W**omen's **A**rmy **C**orps, *or by combining initial letters or parts of a series of words, such as* **radar** *for* **ra**dio **d**etecting **and** **r**anging (Webster's Dictionary). Acronyms are very common in science writing and are used to communicate names of organizations, processes, and concepts. Every field of science has its own acronyms that are often specific to that field. This can result in a lack of clear communication if your audience is not familiar with your field. Acronyms can be confusing even to the author after some time has elapsed! The use of acronyms should be minimized to avoid confusion, and if they are used, they should always be defined the first time they are used.

A jargon-filled sentence full of acronyms and unexplained terms might be:

In years of high precipitation, conductivity of the Chesapeake reduces and loads of TP, TN, and siliciclastic sediments increase. These changes result in HABs as well as reduced DO and K_d which result in reduced growth and spatial extent of SAV.

The translated version, although slightly longer, is much more understandable:

In high rainfall years, more freshwater enters Chesapeake Bay, bringing with it large amounts of nutrients (mainly nitrogen and phosphorus) and terrestrial sediments. This can lead to harmful algal blooms (HABs) in the water, as well as reduction or even removal of oxygen in deeper water. Algae and sediment in the water reduce the ability of light to pass through the water, and absence of oxygen and light can result in reduction or death of aquatic grasses.

ENGAGING YOUR AUDIENCE—PREPARE FOR AND INVITE QUESTIONS

The most productive type of communication is two-way communication. The exchange of ideas and introduction of new concepts can stimulate collaborations, new research questions, and novel problem-solving approaches. Posters and oral presentations are the most common forms of communication and encourage audience engagement and discussion. Always provide an opportunity for feedback. In a talk, this means leaving time for questions; in a written or web setting, this means providing contact details, or online feedback or assessment forms. Feedback aids in evaluating whether the intended audience is getting the message and allows the communicator to demonstrate proficiency and, most importantly, to learn how to make improvements.

REFERENCES

1. Saxby TA, Dennison WC, & Hoegh-Guldberg O. 2003. Photosynthetic responses of the coral *Montipora digitata* to cold temperature stress. *Marine Ecology Progress Series* 248: 85–97.
2. Jones AB, Carruthers TJB, Pantus FJ, Thomas JE, Saxby TA, & Dennison WC. 2004. *A water quality assessment of the Maryland Coastal Bays including nitrogen source identification using stable isotopes.* Report submitted to the Maryland Coastal Bays Program, Maryland, U.S.A.
3. Thomas JE, & Stevenson JC. 2003. *Hurricane Isabel and sea level rise.* Integration and Application Network newsletter #6, Maryland, U.S.A.

Make everything as simple as possible, but not simpler.
Albert Einstein

3.

How do we communicate science effectively?

Communication design determines the ease and success with which your audience will read, comprehend, and assimilate your work. You want to make it as easy as possible for your audience to understand your story. This chapter will introduce you to some practical fundamentals of communication design, including how to format words and typefaces, image resolution and formats, color modes, and graphic file types to use for different media.

MAKE CONTENT THE FIRST PRIORITY

The most important part of your communication products is the content—the information, message, or story you are trying to share. Even the most polished, professional design is pointless if you do not have adequate content. Everything else is packaging, to make it easy for your audience to understand your work and remember you.

DEVELOP A CONSISTENT STYLE AND FORMAT

Choose a color scheme, typeface, and text style, and use them throughout. Simple typefaces such as Arial, Times New Roman, and Helvetica are easy to read. Using common system typefaces like these also reduces problems, for example, when you send a document to a professional printer, or if you are giving a presentation on somebody else's computer and your chosen typeface is not installed. This can also be an issue when sharing files between PC and Macintosh computers.

Maintaining consistency in formatting throughout your document or presentation, including color schemes and typefaces, will give your audience a sense of continuity and prevent them from having to reorient themselves to each page or slide. You can generally use your software to format Master Pages that will be applied to each page or slide. Consistency is especially important if you are producing different kinds of documents, such as presentations, posters, newsletters, and websites—it will present a polished image and your audience will associate one document with another.

FORMAT YOUR TEXT EFFECTIVELY

The way that type is formatted, styled, and designed is a fundamental part of your communication product. There is a lot more to type than Times New Roman, 12 point! Through careful use of different typefaces, colors, sizes, spacing, and justification, you can emphasize different points of your story, and tailor your message for your audience.

Paragraph and typeface styles are usually definable in your software. Styles apply the same formatting to all paragraphs, all headlines, all captions, etc., to maintain consistency. A style can include all aspects of typeface and paragraph formatting such as typeface color and size, line spacing, spacing before and after paragraphs, and justification.

Spelling

This is, without a doubt, the most important component of typography, and unlike most of the other guidelines for typography, there is only one way to do it right. Always use both a software and a human spell-checker. Nothing looks less professional than spelling errors. Correct grammar and punctuation are also fundamental for professional-looking communication products.

Typeface

There are two broad classes of typefaces—serif and sans serif. Serif typefaces have small strokes at the ends of the characters. Sans serif typefaces lack these strokes. The serifs function to guide the eye along a line of text, which is useful when you have a lot of small, tight text, such as the text of a journal manuscript or newspaper article. Sans serif typefaces are more readable from a distance when there is not a lot of text, and so are useful for headlines, titles, presentations, and posters. Pick one serif typeface and one sans serif typeface per document, and stick with them.

Use **bold**, *italics*, CAPITALS, SMALL CAPS, or contrasting colors to create contrast and emphasis, instead of using lots of different typefaces. Also avoid underlining for emphasis—underlining is a relic from typewriter days. Try to avoid using script or other fancy typefaces, as these typefaces are harder to read, and are generally inappropriate for communicating scientific information.

Serif typefaces	Sans serif typefaces
Garamond	Arial
Palatino	Helvetica
Times New Roman	Verdana

Serif   Sans serif

Serifs are these small strokes at the ends of characters

Sans serif typefaces lack these strokes

Text justification

Left justification of text allows the lines of text to end naturally, leaving varied amounts of white space at the end of the line. This is also called flush left-ragged right alignment. Flush left-ragged right alignment adds extra white space between columns, and at the end of lines of text.

Fully justified text is lined up on both the left and right margins, but with different spacing between words. You can fit more text in fully justified paragraphs, however it can make the document look more crowded.

Flush left-ragged right justification has uneven spacing on the ends of the lines on the right margin, but keeps the spacing between the words the same. Flush left-ragged right justification does not look as tidy as fully justified text, but it adds white space at the ends of each line, which can make the overall document look less crowded.

There are no hard and fast rules about which kind of justification to use. It depends on the overall layout of your document, your audience, how crowded or sparse the text is, and the fonts and margins of the document. Try experimenting with different kinds of justification and use what most effectively communicates your message.

For small amounts of text or short, narrow lines, use flush left-ragged right justification, otherwise the different spacing between words becomes very noticeable and impedes easy reading by creating rivers of white space running down the lines. These rivers of white space can be minimized by using your software's preferences to set the maximum and minimum spacing between words.

Having narrow columns of fully justified text creates unsightly rivers of white space.

Double spaces

Two spaces between one sentence and the next are unnecessary. Inserting two spaces between a period at the end of one sentence and the start of the next sentence was the accepted practice when typewriters were used to produce documents. The characters typed on a typewriter are all the same width, i.e., the character i takes up the same width as the character W (fixed spacing). With fixed spacing, two spaces were needed to make a clear distinction between sentences. With the advent of computer word processors, the spacing allocated for characters became proportional to the width of the individual characters (proportional spacing). Characters that are proportionally spaced are easier to read, and so the double space is no longer necessary.

```
With a fixed-width typeface
such as Courier, each
character takes up the same
amount of horizontal space.
```

With a proportional typeface, such as Cronos, the amount of horizontal space each character takes up is proportional to its actual width.

How?

Typeface size

The text in your document or presentation has to be visible. The appropriate typeface size (or point size) to use is different for different media. One point is approximately 1/72 inch. Use a minimum 18 point (pt) typeface size for the body text of a poster, and much bigger (at least 72 point) for the poster title. For presentations, the minimum typeface size is 12 point for small text such as citations and image credits, 18 point for bullet points, and 36 point for slide titles. For newsletters, books, and brochures, the body text typeface size should be at least 10 point, with a minimum size of 14 point for subheads, and 36 point for the title. Typeface size should also be proportional to the length of your lines of text—use a smaller typeface size for narrow columns of text, and a larger typeface size for wider columns of text. Nobody wants to read a line of text on a poster that is two feet wide, set in 10 point!

Medium	Typeface size
Newsletters/books/brochures	
Citations/image credits	8–9 pt
Body text	10–12 pt
Subheads	14–24 pt
Title	36–60 pt
Posters	
Citations/image credits	12–14 pt
Body text	18–24 pt
Subheads	30–48 pt
Title	72–96 pt
Presentations	
Citations/image credits	12–14 pt
Body text/dot points	18–24 pt
Slide titles	36–44 pt

Orphans and widows

Orphans and widows are words, short phrases and lines of text that get separated from the rest of the paragraph, column or page of text to which they belong.

Orphans occur at the bottom of a paragraph, column or page, and widows occur at the top. Orphans and widows leave your reader 'hanging', they compromise the continuity of your document and make your overall layout look messy.

You can use your software's preferences to avoid orphans and widows, by specifying how many lines at the beginning and end of each Alternatively, you can change paragraph must stay together (usually two or three lines). You can also choose to keep headings or subheadings together with the text that follows them, to ensure that a heading does not end up by itself on the bottom of a column or page.

the character spacing (tracking) to reduce or increase the number of lines in a paragraph, to ensure that a word or line does not run over to the next paragraph, column or page. You can also use hyphenation to fine-tune the space that your text takes up.

Orphans and widows separate pieces of text that should stay together. By adjusting the column lengths and character tracking, orphans and widows can be eliminated.

Orphans and widows are words, short phrases and lines of text that get separated from the rest of the paragraph, column or page of text to which they belong.

Orphans occur at the bottom of a paragraph, column or page, and widows occur at the top. Orphans and widows leave your reader 'hanging', they compromise the continuity of your document and make your overall layout look messy.

You can use your software's preferences to avoid orphans and widows, by specifying how many lines at the beginning and end of each paragraph must stay together (usually two or three lines). You can also choose to keep headings or subheadings together with the text that follows them, to ensure that a heading does not end up by itself on the bottom of a column or page.

Alternatively, you can change the character spacing (tracking) to reduce or increase the number of lines in a paragraph, to ensure that a word or line does not run over to the next paragraph, column or page. You can also use hyphenation to fine-tune the space that your text takes up.

Hyphens and dashes

Hyphens (-) are used between hyphenated words, e.g., *professional-looking*, or to break up long words at the end of a line. En dashes (—) are wider than hyphens, and are one en in width. An en is a unit of typographic measurement, and it is equivalent to half the size in points of the typeface you are using, e.g., in a 10 point typeface, an en is 5 points wide, one point being approximately $\frac{1}{72}$ inch. En dashes are used to depict a range, e.g., *0—5 mg l^{-1} of dissolved oxygen*. Em dashes (—) are the widest type of dash, and are one em in width. An em is double the width of an en. Em dashes are usually used in place of colons or parentheses to indicate a break or pause in the sentence—but remember that there should not be a space before or after hyphens, en dashes, or em dashes. Hyphenated words can slow down reading, so try to keep hyphenation to a minimum. You can usually turn off the automatic hyphenation function in your word-processing or desktop design software. Do not hyphenate proper names and never hyphenate the last word of a paragraph.

Other typographic tips

- Write out the numbers zero to nine in words, but use numerals for numbers 10 and higher.
- If the number is at the beginning of a sentence, always write it out in full, including any units that follow it, for example, instead of:

  71% of the earth's surface is covered with water.

  write:

  Seventy-one percent of the earth's surface is covered with water.

  Or you can rearrange the sentence so that the number is not at the beginning:

  Water covers 71% of the earth's surface.

- Avoid Capitalizing The First Letter Of Every Word, It Makes Headlines And Sentences Difficult To Read. Use a capital at the beginning of a headline or sentence, then use lower case letters for subsequent words unless there is a word that requires capitalization, such as the name of an organization.
- AVOID USING ALL CAPITALS FOR ANYTHING OTHER THAN A SHORT HEADLINE. ALL CAPITALS MAKES TEXT HARD TO READ. IF YOU MUST USE CAPITALS, USE SMALL CAPS—THEY READ BETTER THAN REGULAR CAPITALS.
- Use proper curly typographer's quotation marks when using quotation marks and apostrophes (" " ; ' '). Most software supports curly quotes, and you can usually turn them on or off by using the *Smart punctuation* or *Smart quotes* function. Use the prime characters when you want to depict measurements of feet or minutes (′ or '), and inches or seconds (″ or ").
- When using bulleted lists, make sure the bullet or other symbol at the beginning of the text is 'hanging' outside the text, as in the bullet points above, instead of being aligned with the lines of text below it, as in this bullet point. Using 'hanging' indents improves readability.

USE COLOR, BUT USE IT JUDICIOUSLY

One of the characteristics of expanded science communication compared with more traditional science writing in peer-reviewed journals is the more extensive use of color in science communication products. Color can be a powerful and useful tool, but it can be distracting and confusing if not used judiciously.

Using inappropriate colors can make your poster, presentation, or document difficult to look at and alienate your audience. Contrasting, complementary colors are usually the easiest combination of colors to read. Two shades of the same color also work well (e.g., light and dark blue). Colors that are similar to each other do not have enough contrast—they will blend together and be hard to discern from each other. Use another contrasting color to emphasize important points.

In general, for most media including posters, presentations, and websites, the most effective color combinations involve dark-colored text on top of a light-colored background, not the other way around. Looking at a dark background is tiring to the eyes and can make the text difficult to read. A white background is the most effective for websites. A white background can be used for presentations, but white can sometimes look 'dirty' when projected onto a screen.

Some color combinations are particularly difficult to resolve. People with red-green color blindness have difficulty distinguishing between red and green, and even people with normal color vision find red and green combinations hard to read. Another consideration is the potential for black-and-white reproduction of color materials. In both cases, shading and symbol choices can aid in interpretation, even in the absence of color.

Similar colors are hard to distinguish from one another because there is very little contrast.

Two shades of the same color (beige and brown) provide effective contrast. A different color can be used for emphasis.

Some color combinations are very difficult to read. Some people find red and green particularly difficult to distinguish.

This is another example of using two shades of the same color (blue). A contrasting color can be used for emphasis.

Red and green are very powerful colors, and can imply a subjective value, such as good or bad. Use red and green judiciously.

Yellow and blue are complementary colors, providing contrast. Gray is a useful neutral color to use for emphasis.

These are some examples of different color combinations. Contrast between text and background is important, and care must be taken to use colors such as red and green judiciously. Colors such as blue, yellow, orange, brown, black, gray, and white are neutral, in that they do not have a subjective value associated with them as red and green often do.

Sample background colors   **Sample text colors**

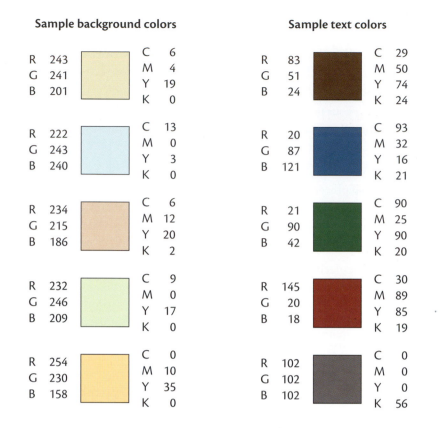

R	243		C	6		
G	241		M	4		
B	201		Y	19		
			K	0		

Some sample colors for background and text, along with their respective settings for both RGB (Red, Green, Blue) and CMYK (Cyan, Magenta, Yellow, Black) color.

RGB or CMYK?

The *RGB* (Red, Green, Blue) color scheme is used when colors are displayed as light, such as on a computer monitor or television screen. RGB is known as an additive color system—when the red, green, and blue are added together, the result is white. All website documents and images are made using RGB color format.

The *CMYK* (Cyan, Magenta, Yellow, Black) color scheme is used when colors are displayed as ink—all printed documents. CMYK is known as a subtractive color system—white is the absence of color, while black is all the colors added together. Actually, you never get white, you just get the color of the paper or media you are printing on—there is no such thing as white ink, unless you use spot color (see below). When creating documents that will end up being printed, make sure that everything is in CMYK format—the text and all the images, as well as the document itself. You can change from RGB to CMYK and vice versa using your image-editing software, such as Adobe Photoshop.

Index Color is a subset of RGB color—it uses only 256 colors, compared with 16 million colors used in RGB, which makes the file sizes smaller and ideal for using on websites. However, be aware that when you convert an RGB image to Index Color, those extra 15,999,744 colors are permanently gone. If you convert the Index Color image back to RGB, there will still only be 256 colors in the image.

Spot color refers to a printed color that is not based on CMYK, but instead is a custom mix color ink, such as white, or fluorescent or metallic colors. Spot color is usually used to make sure a professional printer uses exactly the color that the designer intends. The most common of the spot color standards is the Pantone Matching System, or PMS.

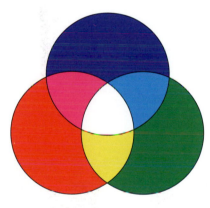

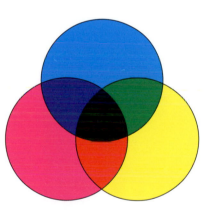

The RGB color system uses three colors of transmitted light to produce different colors. Mixing the three primary colors (red, green, blue) results in white. The absence of light results in black. This illustration is not true RGB, however, as it was necessarily printed as ink. True RGB images can represent a much larger portion of the visible spectrum than CMYK.

The CMYK color system uses four colors of ink to produce different colors. Mixing the three primary colors (cyan, magenta, yellow) results in black. The absence of ink results in white (or the color of the background paper or other medium).

How?

USE THE RIGHT RESOLUTION FOR YOUR GRAPHICS

Resolution refers to the number of dots of ink or electronic pixels, arranged in a grid, that make up an image. Resolution is measured as the number of dots or pixels per inch of space—dots per inch (DPI), pixels per inch (PPI) or even lines per inch (LPI). The higher the density of dots or pixels, the higher the resolution—and the bigger the file size. An image with higher resolution (more dots or pixels per inch) will be clearer (i.e., easier to resolve) than the same image with a lower resolution, even if the two images are the same *physical* size (i.e., take up the same amount of space on a page). Computer monitors can only display up to 96 dots per inch, so all images for websites and on-screen presentations do not need to be any higher resolution than this—the extra pixels or dots are wasted, and will increase your file size unnecessarily. However, please see the section below ('Does resolution matter?'). When you change the physical size of an image, it does not create any more dots or pixels, it just expands or reduces the size of the existing dots and pixels. This can result in an unclear, blurry or 'pixelated' image—if you can see the individual pixels or dots, the resolution is too low. For example, if you have a photo that is three inches long by three inches wide, at 96 DPI, it will look perfectly fine on your computer monitor. If you then double the physical size of the image, to six inches by six inches, the resolution will correspondingly halve to 48 DPI—and your image suddenly looks very blurry on your screen.

Laser printers can print high resolution images—most laser printers print up to 300 DPI, and professional printers can go much higher than that. Always keep an original copy of any image at the highest resolution available, and always work on copies of the original, not on the original file itself.

The image on the left has a relatively low resolution—50 DPI. When printed on a 300 DPI laser printer, the printer 'fills in' the missing dots and creates an unclear and blurry image.

The image on the right has a relatively high resolution—300 DPI. This is the most common resolution used for professional printing, and the image looks clear and sharp.

Does resolution matter?

In short, theoretically it does not matter for electronic publication formats (i.e., websites and presentations), but it does matter for print formats (i.e., posters and newsletters). In reality, the answer is a little more complicated, mostly due to the way that Microsoft PowerPoint sizes images.

A digital image has three parameters that are used to describe its 'size'—pixel dimensions (or actual pixels), document size (or print size), and resolution. As far as the quality of the image goes, the only one of these parameters that really matters is pixel dimensions. This describes the number of actual pixels that make up the image. For example, an image that has pixel dimensions of 2,048 x 3,072 pixels, has a total of 6,291,456 pixels (2,048 multiplied by 3,072). So this image may be set to a document size of 1″ x 1.5″ at a resolution of 2,048 DPI (dots per inch, or pixels per inch) or it may be proportionally scaled to 6.827″ x 10.24″ at 300 DPI. In both cases, it will still be 2,048 x 3,072 pixels in dimension (1″ x 2,048 DPI = 2,048 pixels and 1.5″ x 2,048 DPI = 3,072 pixels, or 6.827″ x 300 DPI = 2,048 pixels and 10.24″ x 300 DPI = 3,072 pixels). In both cases, the quality will appear the same on screen. Only when printing the image will the resolution (DPI) setting have an impact on the view image quality. And, since the pixel dimensions divided by document size equals the resolution, we can effectively describe the image with only two of these parameters.

When scanning images, try this exercise—scan a 35 mm slide at 4,000 DPI and view it using Adobe Photoshop. If you view Actual Pixels (pixel dimensions), the image will be enormous and you will have to scroll around to view the whole image, but if you view Print Size, it will appear the size of the original slide (35 mm). So remember, when you are scanning, that you are scanning a resolution at the current size of the object. When a 35 mm slide scanned at 4,000 DPI is converted to a useful print size (e.g., 8″ x 12″) it will no longer be 4,000 DPI, but closer to 500 DPI. The number of physical pixels has not changed, they are simply taking up more space.

Document size and resolution are settings which software embeds into the image to tell a computer or printer how to display or print the image. They do not affect the quality of the image in any way—remember, it is only the physical dimensions in pixels that affect the information in the image. In the digital medium (i.e., on-screen), resolution is not relevant since pixels cannot be compressed or expanded. The only thing that will determine the quality of an image on-screen is the pixel dimensions, and the only way to change the pixel dimensions is to resample the image, thereby discarding pixels or adding new ones using a mathematical algorithm. Both increasing and even decreasing the number of pixels will reduce the quality of an image, although typically increasing the size has a worse effect on quality. Printing an image is where resolution comes into effect. Remembering that in effect resolution (as well as print size) do not actually exist as real parameters, the resolution setting within the image simply tells the printer how to space and size the pixels.

Screens are typically between 72 and 96 DPI, but this is controlled by a combination of the so-called screen resolution and the physical dimensions of your screen. For example, you might have a 19″ monitor. Your computer screen probably has a resolution of 1,024 x 768 pixels, which sounds an awful lot like pixel dimensions rather than resolution, doesn't it? So, a 19″ screen with a 'resolution' of 1,024 x 768 will display images at … well, this is where it gets confusing because the 19″ measurement is a diagonal calculation of the screen size and the actual horizontal measurement is likely closer to 14″. So, 1,024 x 768 on a 14″ wide screen would be 73 DPI. This number is what will control how large an image is displayed on-screen. So even if you set an image to 1 DPI, it will still display on-screen at 73 DPI and the quality will look exactly the same as if it was set to 4,000 DPI, so long as the actual pixel dimensions have not changed. To prove this to yourself, try changing the resolution of an image in Adobe Photoshop to 1 DPI, being sure that the *Resample Image* option is unchecked so that pixel dimensions do not change. You will notice that at 1 DPI, the document size (if set to pixels) will be the same as the pixel dimensions. You will also notice that on-screen the image still looks the same, but if you try to print it, it will look terrible because it will literally try to spread one pixel over 1″ on the printed page. On screen it will not do this—a pixel will still take up the same amount of space and display at 73 DPI.

Now it gets even more confusing. We now know that resolution makes no difference, but we notice that when we insert an image into Microsoft PowerPoint, it displays the image based on the document size, rather than the pixel dimensions. In contrast, web browsers always display images based on pixel dimensions. The latter means that images will appear quite different in size depending on the screen resolution of the monitor being used. However, Microsoft PowerPoint effectively scales the image to the screen resolution by displaying images on-screen based on document (print) size, rather than the pixel dimensions. The default view in Adobe Photoshop and other dedicated image manipulation programs is pixel dimensions (called *Actual Pixels* in the *View* menu), and so behaves in the same way as a web browser. So for scaling images for use in Microsoft PowerPoint, you need to set the document size (print size) and make sure that the resolution is around 96 DPI—this ensures a sufficient number of pixels in the image (pixel dimensions) to properly display the image using the document size method that Microsoft PowerPoint uses. We said earlier that "In the digital medium (i.e., on-screen), resolution is not relevant since pixels cannot be compressed or expanded", however applications like Microsoft PowerPoint break this rule in a sense by automatically scaling the image to the set document size. Because screen resolution is typically 72–96 DPI, any extra pixels in the image which result in a resolution greater than 72–96 DPI at a given document size (the size you want the image to be displayed at in Microsoft PowerPoint) will be superfluous and Microsoft PowerPoint will have to dynamically rescale the image before displaying it—it will not actually remove these pixels. This is why it is essential to correctly resize an image in Adobe Photoshop first to remove these additional pixels, thereby making the Microsoft PowerPoint file sizes much smaller. This behavior by Microsoft PowerPoint is likely a result of the way its pages are set up in inches and not in pixels like many other on-screen applications such as web browsers.

So, to summarize:

For all print media, you want 300 DPI at the document size (print size) that you require. This is straightforward and simply requires scanning the image at an appropriate resolution so that when scaled to the print size it will be 300 DPI. You can calculate this easily using the *pixel dimensions = document size x resolution* equation.

For on-screen there are two options:

- For the web, just consider pixel dimensions. This, along with the user's screen resolution, will control the size of the image on-screen. N.B., there is nothing you can do about the user's screen resolution, but everything will scale proportionally so your layout will be unaffected.
- For presentations using packages like Microsoft PowerPoint which scale images to the document size, you do need to consider resolution. You can set the document size and the resolution to 96 DPI, with *Resample Image* turned on. Adobe Photoshop will modify the number of pixels in the image by adding or removing them where necessary.

USE THE RIGHT IMAGE TYPES AND FORMATS

There are two main types of images that you can use in your documents—*bitmap* and *vector* images. *Bitmap images* are made up of pixels, and therefore are *resolution-dependent*. Common formats of bitmap images are JPEG, TIFF, GIF, and PNG. EPS files can be either bitmap or vector. Digital cameras and scanners produce bitmap photos and images. Bitmaps are usually scanned, or created in an image-editing software such as Adobe Photoshop or Corel Photo-Paint. You can usually convert an image from one bitmap format to another by using the *Save As* or *Export* functions in your imaging-editing software.

JPEG (Joint Photographic Experts Group) is a very common format used for graphics on the internet and for photographs. JPEG files can be either RGB or CMYK, and are generally small, because of the image compression process. This compression means that a certain amount of information in the image is discarded every time it is saved. If you are working on a single JPEG file, and saving it each time you edit it, the image loses quality every time you save it (using the same file name). If you have an image that you plan on editing, save it as a TIFF file first, do all your editing, then re-save it as a JPEG file when you have finished. You can thereby minimize the amount of compression when you save the file, although this results in increased file size.

TIFF (Tagged Image File Format) is another common bitmap image format, and they can be either RGB or CMYK. This format is very versatile in everything from saving scanned images to photographs. Unlike JPEG files, TIFF files maintain the original image quality, regardless of how many times you save the same document. However, the size of TIFF files can become very large, as they are not compressed in the way that JPEG images are.

GIF (Graphical Interchange Format) is an image format created especially for images intended for display on websites and presentations. GIF files use the Index Color format, a subset of RGB color, which keeps the file sizes very small. GIF files also have the advantage that you can define one of the colors as 'transparent', so you can add any color background behind the image, without creating a white box around the image.

PNG (Portable Network Graphic) is a relatively new image format. Like GIF files, PNG files support transparency. PNG-8 files are comparable in their color range to GIF files (8-bit) and use the same method of transparency, by defining one of the colors as 'transparent'. PNG-24 files have a larger range of color (24-bit) and the transparency is handled by defining the opacity of every single pixel, therefore allowing transparency gradients. PNG files use the RGB color format, and are becoming increasingly used in websites and presentations, however PNG format is not compatible with Internet Explorer browsers at this time.

Most image-editing and vector-drawing software packages have a function where you can *Save for the web* as a GIF or a PNG file, and it will convert the color format and choose the right resolution for web display.

EPS (Encapsulated PostScript) format files can be either bitmap or vector images, and can be either RGB or CMYK. EPS files can also contain any combination of text, images, and graphics. You can save your vector image as an EPS file from Adobe Illustrator, CorelDRAW, and Macromedia FreeHand, which preserves the resolution-independent qualities of the vector image (see below for more information about vector images). This means that the image can be scaled to any size without any loss of quality. EPS is the best file format to use when saving vector images.

Vector images are created using vector-drawing software such as Adobe Illustrator, CorelDRAW, and Macromedia FreeHand. Because vector images are created using mathematical equations (or paths) rather than pixels, they are *resolution-independent*—they can be scaled up to any size and not lose any quality. Vector images are useful when you want to create your own illustrations and graphics, and be able to use them at any size. If you save the vector image as an EPS file (which you can insert into desktop publishing software), it will preserve the resolution-independent qualities of the vector, so you can scale it up to any size. Vector-drawing software also allows you to save or export your vector image as a bitmap image at any resolution that you specify, which can then be inserted into your documents. The process of converting a vector image into a bitmap image is known as *rasterizing*.

Another advantage of vector images is that they can be any shape—saving an image as a bitmap will usually create a white square or rectangle around the image. Bitmap images do not have a transparent background, unless you save the image as a specific format (GIF or PNG, for example).

As a general rule, try to obtain or create vector images wherever possible, to give yourself the maximum number of options possible. Vector EPS files are universal between different computer operating systems and software. If you have a photograph or a scanned image, however, it will always be bitmap, so just keep in mind the size and resolution you will require when you are taking photographs or scanning images.

It is more difficult to convert a bitmap image into a vector image. This process works best on simple line drawings, and you will usually need to do additional editing after the conversion. For more complex graphics, it is usually easier to re-trace the image in vector-drawing software. However, recent advances in vector software technology has greatly improved the automatic tracing capabilities.

It is worth the time investment to create effective graphics in all the different formats you will need for different media, e.g., low resolution RGB PNG files for presentations, and vector or high resolution CMYK EPS or TIFF files for posters and newsletters, etc. These graphics form the backbone of your graphic resources library, and you will use your effective graphics over and over again.

The bitmap image on the left is made up of pixels, which makes the image resolution-dependent.

The vector image on the right is made up of paths connected by anchor points, which makes the image resolution-independent.

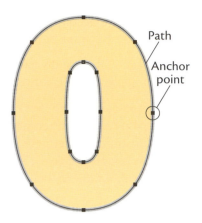

Path

Anchor point

Vector image	JPEG, TIFF, or EPS bitmap image	GIF or PNG bitmap image

Using a vector image means that you can place the image over a background of any color, without creating a white square or rectangle around your image.

Using GIF or PNG images in your website or presentation allows you to make parts of your image transparent, so you can place your image over a background of any color.

Medium	Resolution required	Color format	Image format
Print	300 DPI, or vector	CMYK	EPS (bitmap or vector), TIFF, JPG
Screen	96 DPI	RGB or Index Color	PNG, GIF

FORMAT YOUR VISUAL ELEMENTS EFFECTIVELY

Conceptual diagrams

Ineffective

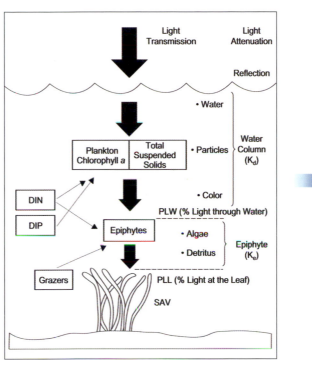

Effective

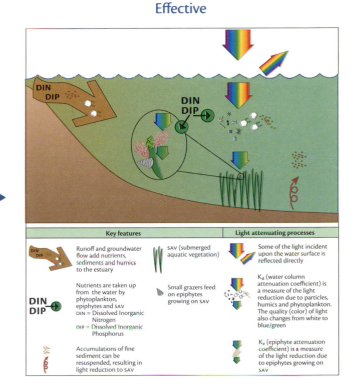

This figure showing light attenuation processes is not self-contained. There are acronyms that are not defined (DIN, DIP, SAV), and processes requiring explanations (K_d, K_e). It also uses text-and-box combinations to depict processes. This figure required half a page of accompanying explanatory text.[1]

This conceptual diagram uses a combination of symbols and color to depict processes. There is a self-contained legend which explains the processes and definitions. Well-constructed conceptual diagrams help your audience to visualize your message.

Satellite photos and maps

Ineffective

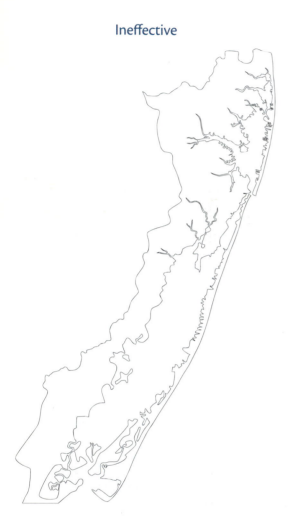

Effective

This line map on the left does not provide enough detail for your reader to orient themselves. It looks like some shoreline, somewhere, but whereabouts in the world, and how big is it?

This figure includes a large-scale locator map (in this case of the Delmarva Peninsula on the U.S. Atlantic coast) to orient the audience. Adding a scale bar and a north arrow provides more orientation information. Local features such as state lines, roads, waterways, and towns provide points of reference and a clear, self-contained legend prevents ambiguity. Distinguishing between the water and land using color or shading further increases the amount of relevant information on the map, without taking up any more space than the original map on the left.

Photographs

Ineffective	Effective

Study site on Amity Banks, eastern Moreton Bay.

Cyanobacterium *Lyngbya majuscula* growing on the seagrass *Zostera capricorni*.

Photos are a powerful way of communicating information and good formatting of photos enhances your message. These photos are different sizes and are not aligned with each other. Putting them next to each other implies a connection between the photos, but that is not made clear. There is no photographer credited, so people will assume that you took the photos yourself.

Crop photos to focus on the message they are portraying. Adjust the size and alignment when using two or more photos to maintain symmetry—here the photos were adjusted to be the same width because they are stacked vertically. Use self-contained captions or text inserts to explain the context of the photos. Use a locator box if one photo is a smaller-scale 'pull-out' of another. Always credit the photographer if it is somebody other than yourself.

CHRIS ROELFSEMA

How?

How to take good photographs

Composition

There are a few very simple rules that you can follow to take a good photograph.

Get in close. Then get in closer!

The first rule for good science photography is to get in close. And when you think you are as close as you could be, get in even closer! If you let your subject fill the frame, you eliminate anything in the background that may detract from your subject. This simplifies the photo, drawing more attention to the subject, and better communicates the story you want to tell with your photograph.

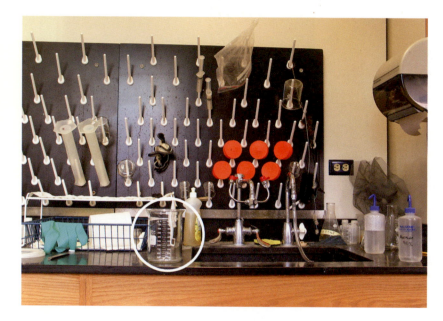 

Which photograph is more effective? The photo on the left where you have to search for the subject of interest, or the photo on the right where the subject matter is clear? Zooming in and cropping the photo is a very effective way to emphasize the point of the photograph.

The rule of thirds

Another common rule of photography is the rule of thirds. The easiest way to think about this is to divide the photograph into thirds, both vertically and horizontally. You then use these guides to position objects along the vertical or horizontal axes, or even better, to place the main subject of an image at the intersecting points created (roughly) when you split the image into thirds. For example, if you are taking a landscape shot, you would place the horizon either one third from the top or the bottom of the frame.

Place the key subject matter of the photograph at the intersections. In this example, the horizon is placed along the bottom third of the frame, and the conifer tree on the left is aligned along the left third of the frame.

Shutter speed

Shutter speed refers to the length of time that the shutter stays open. In lower light conditions, a longer shutter speed allows more light to enter the lens. You can usually hand-hold a camera at shutter speeds down to $\frac{1}{60}$ or $\frac{1}{80}$ of a second. Shutter speeds longer than that, however, require a tripod to prevent the image from being blurry due to the natural unsteadiness of your hands. Having longer shutter speeds is also useful if you want to catch the 'movement' of people or objects in your photograph. Anything that is moving in your photograph while the shutter is open will become blurred and give the impression of movement.

Depth of field

When focusing a camera lens on an object, any other objects at the same distance from the lens will also be in focus. The further an object is from that focal point (closer to, or further from the camera lens), the more out of focus it will be. The depth of field describes how far an object can be in front of, or behind the focal point and still be in focus—the 'in-focus zone'.

The combination of the lens' aperture and focal length control the depth of field. The aperture of the lens refers to the hole that allows light into the camera. This hole opens and closes to let in more or less light (see diagram below). The focal length refers to the 'zoom' of the lens. A wide angle lens may have a 17 mm focal length, while a telephoto lens may have a 300 mm focal length.

The smaller your camera's aperture and the shorter the focal length, the deeper the depth of field is (i.e., the more is in focus). Conversely, the larger the aperture and the longer the focal length, the shallower the depth of field will be (i.e., the less is in focus).

The ratio of the lens' focal length and aperture is called the 'f/stop'. For example, if a 300 mm lens (focal length) has an aperture opening of 50 mm wide, the f/stop will be f/6, because the ratio of $300/50$ equals 6. F/stop values typically range from 2 to 32, depending on the quality of the lens. A high quality lens will have a smaller minimum f/stop value, allowing it to capture more light, which is great for ensuring sharp images in low light conditions. The f/stop value is inversely proportional to the aperture—a small aperture has a high f/stop number (see diagram below). In fact, each increment that the f/stop increases (e.g., f/4 to f/5.6), the incoming light correspondingly halves. This may seem confusing, so a simple rule to follow is: **The higher the f/stop, the greater the range of the in-focus zone.**

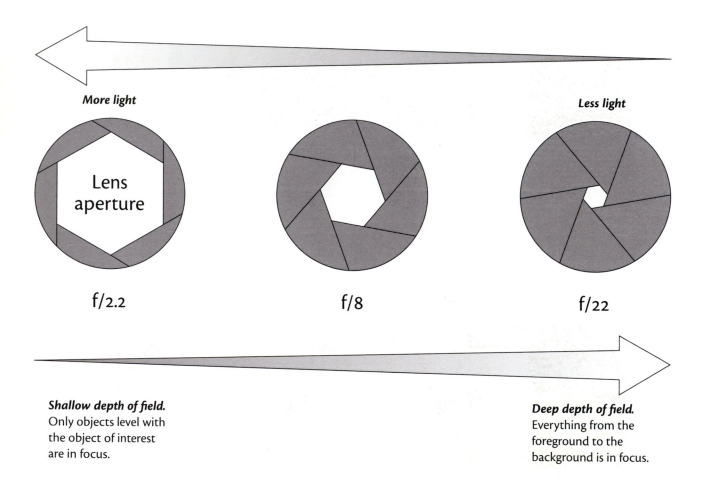

More light   *Less light*

Lens aperture

f/2.2   f/8   f/22

Shallow depth of field.
Only objects level with
the object of interest
are in focus.

Deep depth of field.
Everything from the
foreground to the
background is in focus.

A shallow depth of field (low f/stop) is perfect for shots where you only want a single object in focus, such as a portrait of a person. The object you focus on will be clear and sharp, and the background will be out of focus and blurry. A deep depth of field (high f/stop) is great for when you want to capture everything in a photo, such as in a photo of a landscape. Everything from the background to the foreground will be in focus.

How?

The above photos were taken with differing f/stop values to change the depth of field. The top photo was taken at f/5.6. This low f/stop results in a shallow depth of field, which is evidenced by the apple being the only object in focus. The middle photo was taken at f/27. This medium f/stop results in both the apple and pear being in focus. The bottom photo was taken at f/45. This high f/stop maximizes the depth of field, so that all three objects are in focus. All photos were taken with a lens with 450 mm focal length.

Lighting

Lighting is critical to creating an effective image and properly documenting your science. If your photo is to be taken outside, consider the time of day—early morning or late afternoon is the best time to take many photographs due to the warmth of the light. However, the longer shadows may cause problems if you are trying to capture the intricacies of a complex experimental setup.

Consider the use of a flash outdoors even in sunny conditions to eliminate unwanted or distracting shadows. However, be aware that the range of your flash is quite limited. Turn your flash off in situations when the subject of your photo is too far away to be illuminated by the flash, such as a landscape photo of a mountain. If you are shooting directly through glass or underwater, use an external flash with a cord so you can change the angle of the flash to prevent reflection—a regular flash is at the same angle to the object of interest as the lens, so any reflection from the flash on the glass will be in the photo. A good quality external flash also allows for changing the intensity of the flash. When combined with a digital camera, you have the ideal tool for tweaking flash intensity and light angle to get the best photo. A second pair of hands or a tripod for the flash come in very handy. If you do not have an external flash, turn off the camera's flash and use a tripod. Also, be aware of reflections from room lighting which can be especially problematic with aquaria.

The photo above was taken through the glass of an aquarium using the flash attached to the camera. The reflection of the flash on the glass is clearly visible on the right-hand side of the photo. The photo below is the same photo taken using a tripod and no flash. The elements in this photo are clearer and not obscured by the flash's reflection.

Cloudy days can be great for close-up shots as the light is even and often free of shadows or highlights, however, landscapes will look drab and gray. Using a polarizing filter can help to eliminate glare from the sun and is especially useful for shots looking into the water.

Digital image quality

When shooting with a digital camera, make sure that you are using the highest image quality available with your camera. The more information that is captured by your camera's sensor, the bigger you can print your photographs while still retaining print-quality resolution. This is especially important for posters and larger print media.

Be aware of the ISO setting that you are using. Digital cameras behave the same way as film, in that the higher the ISO, the 'faster' the film, and the grainier the image. However, it is better to get the shot than not, so in low light situations increase the ISO to whatever level you need, but be aware that this will reduce the quality of the photograph. A better solution is to use a lower ISO and a longer shutter speed, with a tripod to stabilize the camera.

Organization

It is important to plan photo sessions along with your research. Often, you will only get one opportunity, so be prepared and consider what photos you may want for reporting your results. It is often too late once you are writing up your results to work out the photo you need. Organize to have a camera close at hand during your experiments, and if necessary, organize for a colleague to take photographs for you. Consider taking these photos where appropriate to your research.

How?

Location

Where are you conducting your research? Are you sampling corals on a tropical island or studying interspecies relationships in a temperate rainforest? Landscape-scale photographs will quickly familiarize your audience with the location of your research. They can also create interest at a local level when the audience recognizes their local environment. Aerial shots of the area are an excellent tool for many projects as it helps to put your site into perspective with the region. Remember that you may not actually have to go to the expense and time of doing this yourself—many resources are available on the internet.

Aerial photos such as this one of Ocean City, Maryland, provide an excellent site-scale reference.

Species

Take a good close-up of the species you are studying, preferably within its natural habitat. This will give your audience a good idea of what your organism looks like, and what kind of environment it inhabits.

A close-up photo, such as this one of the seagrass *Thalassia testudinum*, is an excellent way to introduce your audience to your species of interest.

Experimental design and set-up

Good shots of your experimental set-up, often together with annotations, can be used very effectively to explain how it all worked.

KATHARINA ENGELHARDT

This photograph is a great example of using a photo to depict your experimental set-up. In this case, it shows the arrangement of the experimental mesocosms.

Methods

Taking photos of the experimental methods you use, or setting up shots to demonstrate your methods can be a great help in explaining your methodology.

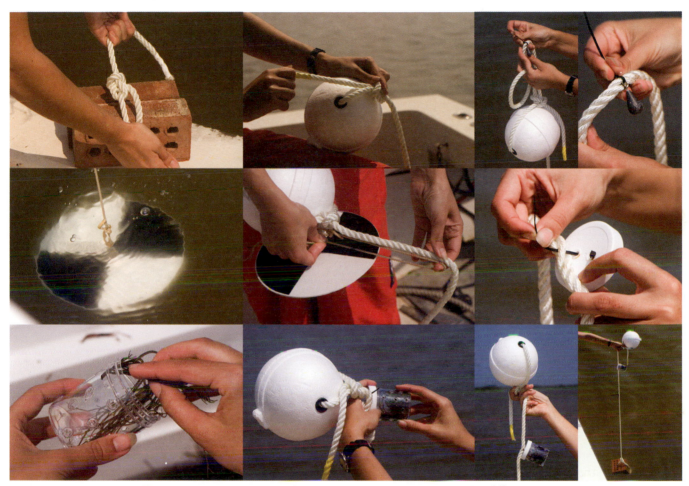

This sequence of photos effectively demonstrates a methodology (preparation of a macroalgal δ^{15}N incubation rig). From top left: securing rope to bricks for anchor; tying on the buoy; cable-tying a sinker part-way down the rope to keep the chamber under water; the sinker attached to the rope; measuring the Secchi disk depth; finding the point on the rope that is half the Secchi disk depth to standardize the light regime between sites; cable-tying the chamber lid to the rope at 'half Secchi disk depth'; placing approximately 5 g (wet weight) of macroalgae into the perforated chamber; screwing the chamber into its lid; the finished rig; and deploying the rig at a GIS-located site.

Any visual results

What happened during your experiments? Did the subjects respond well/badly? Were there any visible differences between treatments? A visual response can be more powerful than a graph or table of numbers to show your audience the results of your experiments. Whether you expect a visual change or not, it is a great idea to get the before shots anyway, just in case.

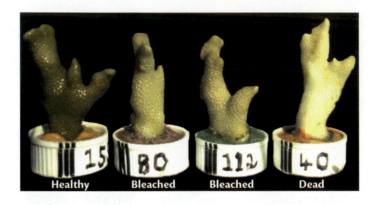

This photo taken at the end of an experiment graphically illustrates the experimental results. The results show corals exposed to different temperature treatments, resulting in different levels of coral bleaching.

Photo organization

A common fault is that people do not take care of their photos. Slides need to be carefully stored in a dry place to minimize contamination by mold, fungus, and dust. Slides and photographs also seem to be easily misplaced amongst the clutter of a scientist's office. Try to keep them in one place.

Take the time to label your slides and photographs. It is inevitable that the memory fades with time. The photographer may also move on to another job and with them goes the information about the photo. Try to record the name of the photographer, the date, the location, who is in the photograph, what is happening in the photograph, and file this with the photograph in a safe place. Setting up and maintaining a database is a great idea to ensure proper recording of information.

Graphs

Ineffective

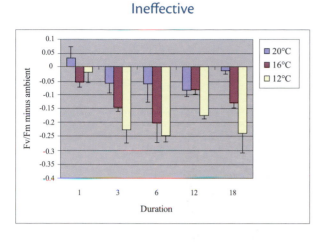

Effective

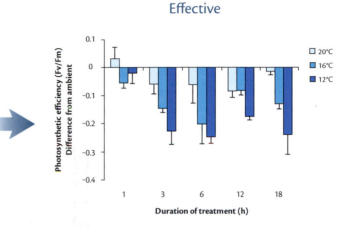

This is a typical graph output from Microsoft Excel, using the default settings. The gray background with the gridlines is distracting and reduces the impact of the data. There are also too many boxes—around the graph itself, around the legend, and then around the whole graphic. The only lines should be the two axes. Also make sure that your axes are clearly labeled with both what is being shown, and the units of measurement. While the label 'Fv/Fm minus ambient' might be completely understandable to you, it is too ambiguous for a general audience. The x-axis 'Duration' also needs more explanation and the units need to be labeled. The typeface used is Times New Roman, a serif typeface. Graphs are generally printed at a small size, and a sans serif typeface is more legible at smaller sizes. Finally, the color scheme used is meaningless. Color is a powerful tool for communicating information, and should be used to enhance your message.

This is the same graph after some minor formatting in Microsoft Excel. Simplify the graph by removing the boxes around the graph, the legend, and the whole chart, as well as the gridlines. The number of tick marks on the y-axis has been reduced by half to remove unnecessary clutter. Both axes need to be clearly labeled with what they are depicting, as well as the units in which they are measured. Use a sans serif typeface to enhance readability at small sizes. Color can convey information—here the increasingly cooler temperatures of the experiment are depicted by increasingly blue (cooler) colors.

How?

Tables

Ineffective

Reporting region	Water Quality Index	Chlorophyll *a*	δ^{15} Nitrogen	Dissolved oxygen	Total nitrogen	Total phosphorus	Secchi depth
Assawoman Bay	0.56	1	0.63	1	0.45	0.06	0.15
Chincoteague Bay	0.42	0.40	0.76	0.87	0.40	0	0.05
Chincoteague Appendix	0.62	0.86	0.80	1	1	0	0.02
Isle of Wight Bay	0.69	1	0	1	1	0.44	0.70
Newport River	0.36	0	0.83	1	0.21	0.07	0
Newport Bay	0.33	0.11	1	0.89	0	0	0
Sinepuxent Bay	0.68	0.88	1	0.96	0.86	0.21	0.22
St. Martin River	0.29	0.45	0.25	1	0.03	0	0

Lines around every cell in a table compartmentalize the flow of information and make it difficult to make comparisons along rows and columns.[3]

Effective

Reporting region	Water Quality Index	Chlorophyll *a*	δ^{15} Nitrogen	Dissolved oxygen	Total nitrogen	Total phosphorus	Secchi depth
Assawoman Bay	0.56	1	0.63	1	0.45	0.06	0.15
Chincoteague Bay	0.42	0.40	0.76	0.87	0.40	0	0.05
Chincoteague Appendix	0.62	0.86	0.80	1	1	0	0.02
Isle of Wight Bay	0.69	1	0	1	1	0.44	0.70
Newport River	0.36	0	0.83	1	0.21	0.07	0
Newport Bay	0.33	0.11	1	0.89	0	0	0
Sinepuxent Bay	0.68	0.88	1	0.96	0.86	0.21	0.22
St. Martin River	0.29	0.45	0.25	1	0.03	0	0

Simplify your tables. Deleting vertical lines, keeping horizontal lines to a minimum, and increasing the spacing between rows makes the table much more readable.

REFERENCES

1. Batiuk RA, Bergstrom P, Kemp M, Koch E, Murray L, Stevenson JC, Bartleson R, Carter V, Rybicki NB, Landwehr JM, Gallegos C, Karrh L, Naylor M, Wilcox D, Moore KA, Ailstock S, & Teichberg M. 2000. *Chesapeake Bay submerged aquatic vegetation water quality and habitat-based requirements and restoration targets: A second technical synthesis.* U.S. EPA and Chesapeake Bay Program, Annapolis, Maryland, U.S.A.
2. Saxby TA, Dennison WC, & Hoegh-Guldberg O. 2003. Photosynthetic responses of the coral *Montipora digitata* to cold temperature stress. *Marine Ecology Progress Series* 248: 85–97.
3. Jones AB, Carruthers TJB, Pantus FJ, Thomas JE, Saxby TA, & Dennison WC. 2004. *A water quality assessment of the Maryland Coastal Bays including nitrogen source identification using stable isotopes.* Report submitted to the Maryland Coastal Bays Program, Maryland, U.S.A.

FURTHER INFORMATION

The books and websites below are current at the time of printing. However, as software is constantly upgraded and changed, these books and websites may also change and be updated. This is not an exhaustive list, but a small sample of useful resources.

Adobe Creative Team. 2005. *Adobe Photoshop CS2 classroom in a book.* Adobe Press, Berkeley, California, U.S.A.

Blatner D, & Chavez C. 2005. *Adobe Photoshop CS2 tips and tricks.* Adobe Press, Berkeley, California, U.S.A.

Bringhurst R. 2004. *The elements of typographic style.* Hartley and Marks Publishers, Point Roberts, Washington, U.S.A.

Felici J. 2002. *The complete manual of typography.* Adobe Press, Berkeley, California, U.S.A.

Kelby S, & Nelson F. 2005. *Photoshop CS2 killer tips.* Peachpit Press, Berkeley, California, U.S.A.

Integration and Application Network   *www.ian.umces.edu*

Adobe Systems Incorporated   *www.adobe.com*

Designer Today Graphic Design Magazine   *www.designertoday.com*

My Design Primer   *www.mydesignprimer.com*

Color Matters   *www.colormatters.com*

Typography 1st   *www.typography-1st.com*

Typophile   *www.typophile.com*

4.

Conceptual diagrams

Science communication is an essential component of environmental problem solving. Effective scientific communication requires synthesis, visualization, and appropriate context. Conceptual diagrams, or 'thought drawings', are an excellent means of providing these requirements. A conceptual diagram uses symbols to convey the essential attributes of a system. There are four important reasons for using conceptual diagrams:

- to clarify thinking and avoid ambiguities;
- to provide a unique communication interface between scientific disciplines or between scientists and non-scientists;
- to identify gaps in research and knowledge, establish priorities, and solicit an agreed synthesis;
- to better define scales of processes and linkages within habitats and communities.

CONCEPTUAL DIAGRAMS ARE 'THOUGHT DRAWINGS'

'Concept' comes from the Latin 'conceptus', and means *something conceived; or something formed in the mind* (Webster's Dictionary). 'Diagram' comes from the Greek 'diagramma', and means *a figure worked out with lines; or a plan, sketch, drawing, or outline designed to demonstrate or explain how something works or to clarify the relationship between the parts of a whole* (Webster's Dictionary).

Based on the idea that a diagram can be worth 10,000 words,[1] conceptual diagrams provide diagrammatic representations of ecosystems in which key features and major impacts can be illustrated. Good conceptual diagrams are used extensively, and have many lives in many places.

Textbooks are replete with versions of the better conceptual diagrams. An example of a popular conceptual diagram used in science is the 'z scheme' of photosynthesis, depicting the electron flow in the light reactions of photosynthesis, with energy level of the electron used as the *y*-axis, and time as the *x*-axis. This basic scheme is used so widely that it is named after the diagram's visual appearance, which roughly approximates the letter 'z'.

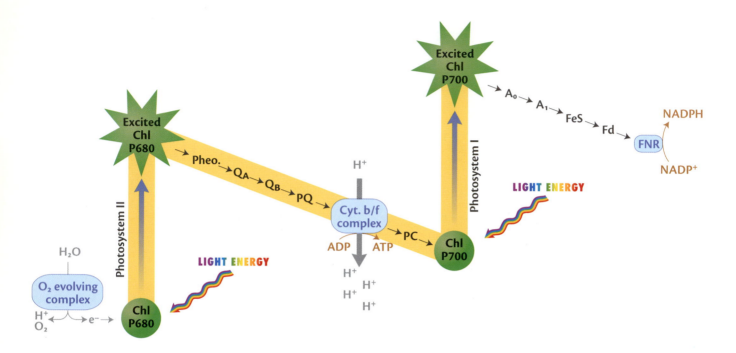

An example of a popular textbook conceptual diagram—the 'z scheme' of photosynthesis.

Another popular conceptual diagram is the depiction of the process of plate tectonics. A cross-section of the earth's lithosphere is used to illustrate the processes of tectonic plate formation, intersection, and loss. Volcanoes, mountain ridges, oceanic trenches and rift zones are depicted at the tectonic plate boundaries.

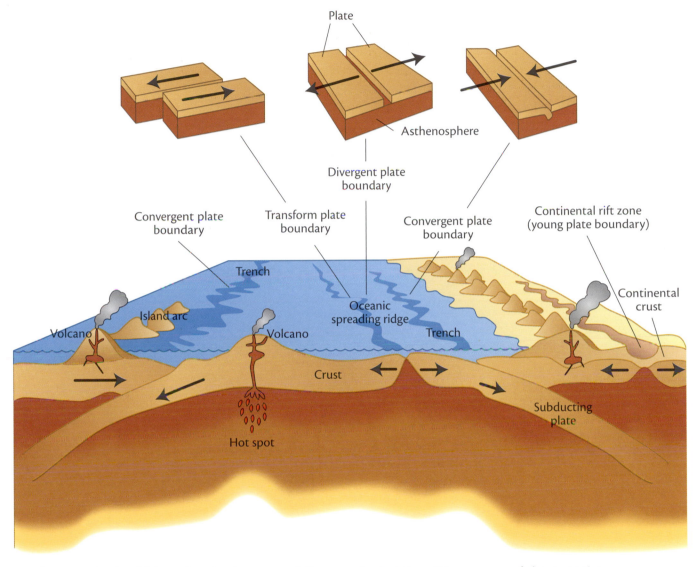

Another example of a widely used textbook conceptual diagram—the depiction of the processes of plate tectonics. Cross-sections of the earth's lithosphere are used to illustrate the processes of plate tectonic activity.[2]

SYMBOLS FORM A VISUAL LANGUAGE

One of the key aspects of conceptual diagrams is the use of symbols. Symbols are one of the most ancient forms of human communication and remain a common feature of everyday life. They are universally used for depicting messages that can transcend culture, language, and time. The size, shape, color, and position of symbols all convey meaningful information, and when arranged into a diagram, they can augment or replace words. The use of symbols to construct conceptual diagrams can be an effective tool for science communication and problem solving.

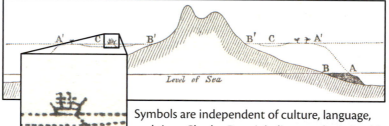

Symbols are used in mathematics (e.g., π), music (e.g.,), weather (e.g.,), religion (e.g.,), and organizations (e.g.,). Symbols can be universally understood, are language independent, and are an important feature of everyday life (e.g.,). Scale can be important in the use of symbols—the size of the symbol can represent relative importance (e.g., vs.). Color and shape of symbols are also important, as in nautical navigation when weather conditions may partially obscure the navigational aids—the combination of shape and color ensures the message is still clear (e.g., 5 vs. 4).

Symbols are independent of culture, language, and time. Charles Darwin's diagram depicting his theory of coral reef formation (above) was first published in 1842, and has been reproduced thousands of times since.[3] Yet the depiction of a sailboat is clear, as is the image of a sailboat created centuries ago in a cave painting by an indigenous Australian (right).

NORM DUKE

The use of symbols in conceptual diagrams necessitates the creation of a comprehensive legend explaining the symbols. Inclusion of a legend makes the conceptual diagram self-contained, with no need to read an accompanying explanation. In this way, an unambiguous interpretation of the symbols is developed, and allows for the inclusion of references, as footnotes or author citations, to support each of the symbols or concepts introduced in the conceptual diagram.

WHY USE CONCEPTUAL DIAGRAMS?

Conceptual diagrams provide synthesis, visualization, and context. They are an effective tool to communicate complex messages in an accessible, informative, and visually interesting manner. Conceptual diagrams help to clarify thinking—words can be ambiguous but an image commits to the message being portrayed. Often, a topic that is poorly understood can be written about, but may not lend itself to drawing a diagram. The act of attempting to draw a diagram will often force scientists to crystallize their thoughts. Visual communication tools such as conceptual diagrams are much more effective at communicating large amounts of complex information than text alone.

Conceptual diagrams are useful for scientific integration and application. They can identify gaps, establish priorities, and solicit an agreed synthesis. Conceptual diagrams can be used in proposals as a way of identifying research priorities, in scientific syntheses to integrate research findings, in establishing the relationships between indicators used in monitoring programs and the parameters they represent, and in management applications by identifying the various problems and proposed actions addressing the problems.

The creation of conceptual diagrams facilitates communication between scientists, resource managers, and non-scientists by assisting in both one-way (idea presentation, e.g., presenting syntheses) and two-way (idea development, e.g., soliciting an agreed synthesis) communication. In communication between scientists and the wider community, conceptual diagrams can combine the scientists' current understanding with the community's priorities and environmental values, resulting in a shared vision which is essential in solving environmental problems. By using symbols instead of just words, and presenting a diagram that encapsulates major findings and issues, conceptual diagrams can stimulate a dialogue between scientists and stakeholders.

Once the techniques of producing conceptual diagrams have been mastered, they can be produced in real time to synthesize key messages or consensus at workshops and meetings.

Good conceptual diagrams have many lives in many places. They serve to highlight essential ecosystem features and interactions. They can depict biota and processes at different scales, from global to local scales and from organismal to subcellular or even subatomic scales, and can evolve to capture an increasing understanding of a system.

Conceptual diagrams can be used in a variety of publications, including traditional, peer-reviewed scientific papers (in color or black-and-white), and in science communication products such as presentations, posters, newsletters, books, and websites. In products such as these, conceptual diagrams provide a communication interface between scientists, resource managers, community groups, and the public.

Generation of a conceptual diagram involves identifying the message, identifying the audience, and then listing the elements and processes you wish to depict. You can then experiment with different ways to visualize these elements and processes. A good conceptual diagram usually requires five to 10 iterations to fully capture the required messages.

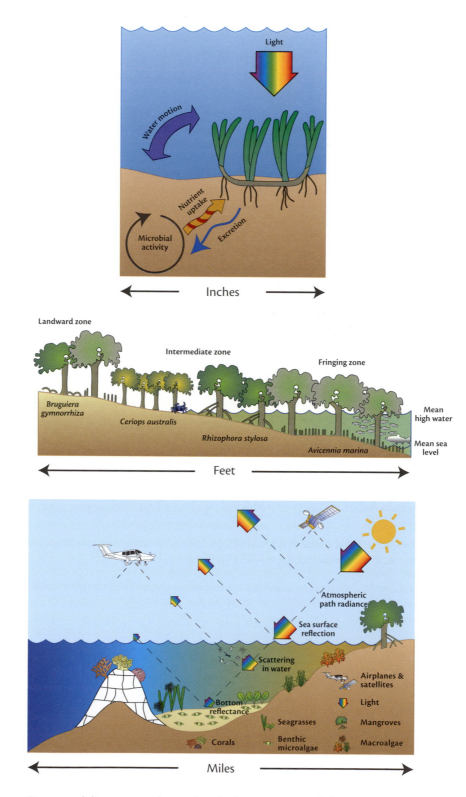

Conceptual diagrams can be used to depict processes at different spatial scales, such as these three diagrams that depict processes at inch, foot, and mile scales.

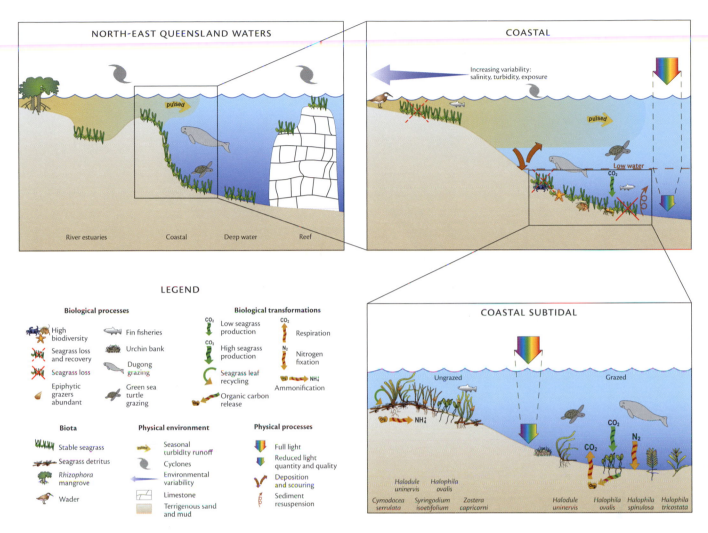

NORTH-EAST QUEENSLAND WATERS

COASTAL

COASTAL SUTIDAL

Conceptual diagrams can be nested, to show processes at increasingly smaller spatial scales, as with these three diagrams depicting processes and biota in coastal waters in Queensland, Australia.[4]

Chincoteague Bay benthic animal communities were intact 🐟 although forage fish abundance has declined ⬭. The macroalga *Chaetomorpha* 🌿 was most abundant in the eastern portion of Chincoteague Bay, where it was associated with seagrass 🌱.

North Chincoteague Bay had moderate to poor water quality ▨ adjacent to the mouth of Newport Bay, and good water quality ▩ elsewhere. Nitrogen concentrations ranged from very low Ⓝ in the western portions to moderate Ⓝ in the east, while phosphorus concentrations ranged from high Ⓟ in western areas to moderate Ⓟ in the east. Chlorophyll (algal) concentrations were very low 🔵 in north Chincoteague Bay. Water clarity (Secchi depth) was good ▾ in western areas, and very good ▾ in the east. Brown tides 🟤 occurred in the north-western part of Chincoteague Bay, associated with the poorer water quality in this region.

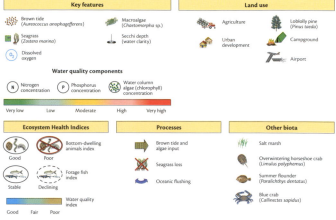

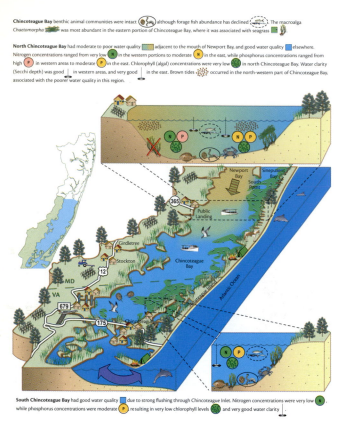

South Chincoteague Bay had good water quality ▩ due to strong flushing through Chincoteague Inlet. Nitrogen concentrations were very low Ⓝ while phosphorus concentrations were moderate Ⓟ resulting in very low chlorophyll levels 🔵 and very good water clarity ▾.

This conceptual diagram was used in a public report to summarize the ecological state of Chincoteague Bay, one of Maryland's Coastal Bays.[5] The self-contained legend (above) complements the use of the symbols in text explaining the diagrams (left). Additionally, the shape of the base of the conceptual diagram mirrors the shape of Chincoteague Bay itself and its watershed.

Conceptual diagrams can be used in science communication products such as newsletters[6] (left), and books[7] (right).

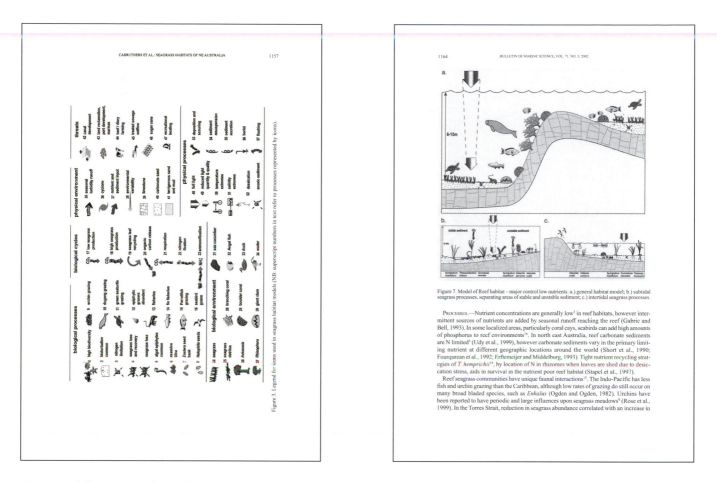

Conceptual diagrams can be used in peer-reviewed scientific journals to facilitate communication between scientists. These diagrams can be just as meaningful when converted from full color to black-and-white or grayscale.[4]

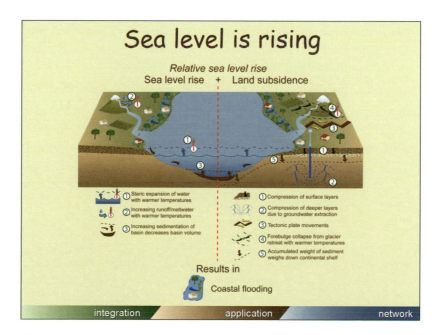

The visual nature of conceptual diagrams make them ideal to use when giving presentations.

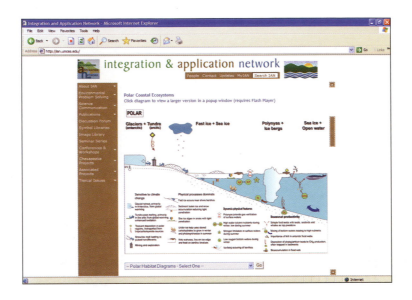

Conceptual diagrams are well-suited for use in websites. They can be used as an interactive index with hyperlinks to further information.

CREATE YOUR OWN CONCEPTUAL DIAGRAMS

Technological advances have now made it possible to generate conceptual diagrams without graphic art training or specialized equipment. Software has developed such that bases and symbols can be added to a diagram with simple 'click and drag' functionality. Rapid generation and iteration can now occur without the intervening step of editing a graphic artist's work. This technological breakthrough will result in a similar transition as that from using typewriters and typists to word processing.

The Integration and Application Network (IAN) has produced a series of scientific symbol libraries, for use with graphic design software. The libraries contain over 1,500 custom-made symbols in 32 categories designed specifically for enhancing science communication skills. Diagrammatic representations of complex processes can be developed easily with minimal graphical skills. Our aim is to make them a standard resource for scientists, resource managers, community groups, and environmentalists worldwide.

The IAN symbol libraries have been downloaded by thousands of people in nearly every country. The libraries are available cost- and royalty-free. Also available for download is a searchable index (PDF) of all the available symbols and an interactive Flash tutorial on how to use the symbols with Adobe Illustrator. The IAN website (*www.ian.umces.edu*) also has discussion forums about environmental problem solving and science communication techniques, as well as an image library for use in science communication publications.

REFERENCES

1. Larkin J, & Simon H. 1987. Why a diagram is (sometimes) worth ten thousand words. *Cognitive Science Journal* 11: 65–99.
2. Vigil, JF from *This Dynamic Planet*, a wall map produced jointly by the U.S. Geological Survey, the Smithsonian Institution, and the U.S. Naval Research Laboratory.
3. Darwin C. 1874. *The structure and distribution of coral reefs (2nd ed.)*. Smith-Elder, London, UK.
4. Carruthers TJB, Dennison WC, Longstaff BJ, Waycott M, Abal EG, McKenzie LJ, & Long WJL. 2002. Seagrass habitats of north eastern Australia: models of key processes and controls. *Bulletin of Marine Science* 71(3): 1153–1169.
5. Wazniak C, Hall M, Cain C, Wilson D, Jesien R, Thomas J, Carruthers T, & Dennison W. 2004. *State of the Maryland Coastal Bays*. Maryland Department of Natural Resources, Maryland Coastal Bays Program, and University of Maryland Center for Environmental Science, Maryland, U.S.A.
6. Chesapeake Bay Program and the Integration and Application Network. 2005. *Chesapeake Bay environmental models*. Integration and Application Network newsletter #11, Maryland, U.S.A.
7. South East Queensland Regional Water Quality Management Strategy Team. 2001. *Discover the waterways of south-east Queensland*. South East Queensland Regional Water Quality Management Strategy, Brisbane, Queensland, Australia.

FURTHER INFORMATION

The books and websites below are current at the time of printing. However, as software is constantly upgraded and changed, these books and websites may also change and be updated. This is not an exhaustive list, but a small sample of useful resources.

Adobe Creative Team. 2005. *Adobe Illustrator CS2 classroom in a book*. Adobe Press, Berkeley, California, U.S.A.

Cross D. 2005. *Illustrator CS2 killer tips*. Peachpit Press, Berkeley, California, U.S.A.

Frutiger A. 1998. *Signs and Symbols: Their design and meaning*. Watson-Guptill Publications, New York, New York, U.S.A.

Karlins D. 2005. *Adobe Illustrator CS2 tips and tricks*. Adobe Press, Berkeley, California, U.S.A.

Steuer S. 2003. *The Adobe Illustrator CS Wow! Book*. Peachpit Press, Berkeley, California, U.S.A.

Integration and Application Network   *www.ian.umces.edu*

Adobe Systems Incorporated   *www.adobe.com*

Designer Today Graphic Design Magazine   *www.designertoday.com*

My Design Primer   *www.mydesignprimer.com*

Color Matters   *www.colormatters.com*

About Graphic Design   *www.graphicdesign.about.com*

Conceptual diagrams

Right and wrong do not exist in graphic design. There is only effective and non-effective communication.

Peter Bilak

5.

Desktop publishing

Desktop publishing is the process of designing and laying out pages using software on your desktop computer. The end products can be anything from a one-page brochure, to a newsletter, poster, or book. Two of the most common software programs used for desktop publishing are Adobe InDesign (formerly PageMaker) and Quark XPress. Your software comes with extensive manuals and help databases, so the purpose of this chapter is not to be a technical guide to desktop publishing software, but instead to give an overview of the general principles of desktop publishing, page design, and layout.

There are many elements to consider when designing a desktop published document, the basics of which are covered in this chapter. However, always remember that your content comes first. Graphic design and desktop publishing may seem intimidating to those who are not professional designers, but remember that the majority of people using these skills have no professional training. Yes, a good 'eye' and a talent for design and color certainly helps, but most of these techniques and principles are skills that can be learned, just like any other skill.

WHAT IS GRAPHIC DESIGN?

Graphic design is the process of combining text and graphics to communicate effectively using a visual medium. Successful graphic design attracts attention, adding value to the message you want to get across. It can enhance readership and readability by simplifying and organizing your project and selectively emphasizing key points.

STEPS IN THE GRAPHIC DESIGN PROCESS

Analyze your audience

Information must be noticed before you can get your message across. It has to stand out from the crowd by being different. Before you decide how to grab your readers' attention, you should consider who will read it and where it will be seen. Your design should be suitable for your audience and appropriate for its environment.

Get to the point

Choose a key message that you want to get across to your audience.

Decide where and how your message will appear

Will it be a printed publication, such as a newsletter or poster, or a presentation or website?

Split your material into logical sections

Organize your text and graphics. Choose what you want to use carefully—pick images and graphics that will get your main message across. Discard anything that might distract the reader or detract from your message.

Draw rough thumbnails to conceptualize your layout options

Rough sketches can help you develop ideas for layouts, try out themes, and brainstorm on how best to present information. Choose an appropriate format and layout for your publication (first impressions are hard to erase).

Once you decide on a theme, make it consistent throughout your document. Select appropriate typefaces, type sizes, type styles, and spacing for your document. Try to use colors, styles, and typefaces that complement each other. Of course, style is very subjective—everyone has their own idea of what looks good, but do not minimize the importance of layout and design. It is often the very first thing that people notice.

Arrange your information to get your message across clearly

Good layouts are easy to follow and provide the reader with clear cues to help them find their way through a publication easily. If the reader has to work to navigate a document, they probably will not read it. Arrange and emphasize your information to make your message as clear as possible. Decide what is most important and position it accordingly. Continue arranging and emphasizing the information until you have included everything. The quality of your layout will determine how quickly your reader will be directed through the publication and how fast they will be able to read it.

Refine and review, review, review

Print small, draft versions of your project on your own printer. Look at your design with a critical eye. Is it eye-catching? Does it look boring? Is it too busy?

In addition to a spell-check, make sure that you proofread! Or even better, get someone else to do it for you. Their fresh eyes will often pick up mistakes and layout problems that you have missed.

THERE ARE SIX PRINCIPLES OF DESIGN

Every design contains certain basic elements or building blocks that you choose to convey your message. How you organize those items on a page will determine the structure of your design. This can affect the overall readability of your project and will determine how well your design communicates the desired message. The following six principles will help you to combine the various elements of design into a good layout for your desktop publishing project.

Balance

Balance refers to the harmonious arrangement of components in a design, or visual equilibrium. When you are arranging elements on a page you should consider their visual weight, determined by their size, darkness or lightness, and the thickness of lines. How can you arrange components so that they relate to each other within the page to achieve visual balance?

There are two different types of balance in page design: symmetrical and asymmetrical. Symmetrical balance is the arrangement of elements so that they are evenly distributed to the left and to the right of center. It is easiest to see symmetry in a perfectly centered composition, or in those with mirror images. Symmetrical balance is the most appropriate to use for traditional and conservative publications that are formal or orderly.

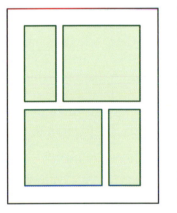 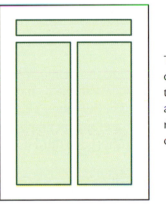

These two mirrored designs have the most complete symmetry possible. The design on the left is symmetrical about both the vertical and horizontal axes, while the design on the right is symmetrical about the vertical axis only.

This poster divides the page into four equal sections. While the content within these four sections is not mirrored, the overall look is very symmetrical and balanced. The different shapes of the graphics are obviously asymmetrical. This concept will be discussed on the following page.

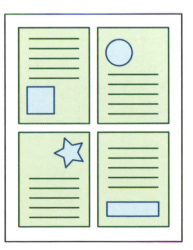

Desktop publishing

Asymmetrical balance is a little more complex and difficult to envisage. It is typically created with an odd or mismatched number of fundamentally different elements to create an off-center design. You can place these objects in a way that will allow objects of different visual weights to balance each other. For example, you might balance a large graphic with a cluster of smaller graphics. Alternatively, you can intentionally avoid balance in your design to create asymmetrical tension.

Layouts that use asymmetry can appear more interesting and dynamic, implying contrast, variety, surprise, movement, and informality. Asymmetrical designs are more appropriate for modern publications.

The design on the left balances one large object with a cluster of smaller objects. In contrast, the design on the right creates asymmetrical tension, with one large object on the left balanced by a single small object on the right.

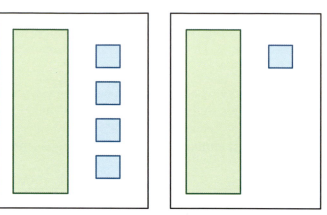

The rule of thirds is a guideline for the placement of elements in a design. Mentally divide the page into thirds vertically and/or horizontally, and place the most important elements at one or more of the four intersections of those lines. You can also arrange areas into bands occupying a third of the space. This rule is not just useful for designing page layout, it can also be applied when you are composing or cropping graphics or photos. Try to place the horizon, or the object of interest (e.g., the eyes of a person in a headshot) one third down the page.

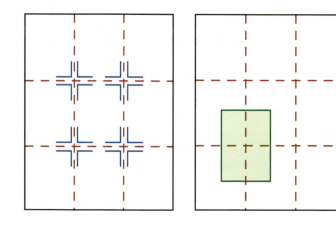

These pages are divided into thirds. The page on the left shows the four intersections where the imaginary grid meets. The page on the right shows the object of interest placed over one of these intersections.

Another trick to balance a page is to place important elements, or the focal point of the design, within the visual center of the page. The visual center is slightly to the right and above the actual center of a page.

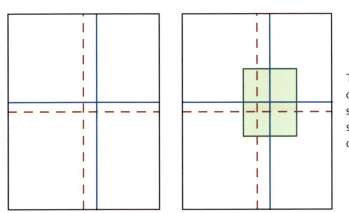

The red dotted lines show the physical center of both of these pages, while the blue lines show the visual center. The page on the right shows an object placed over the visual center of the page.

Proximity or unity

The principles of proximity or unity help to make all the elements look like they belong together. Items that are related should be positioned close together, while items that are not related should be farther apart. This principle helps you to organize your material on the page. It will also give the reader visual cues, so your key message is interpreted quickly and easily.

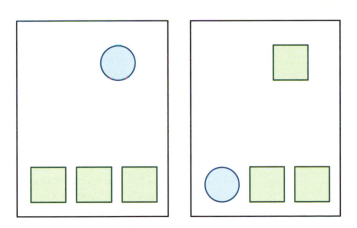

In the page on the left, it is obvious that the three green squares are grouped together, while the blue circle is not related. However, in the page on the right, the circle and the two bottom green boxes belong together, even though the circle is very different to the two squares, while the top green square is not related.

Desktop publishing

Alignment

Alignment can also be used to unify the page so that the elements are a part of a larger whole. The main principle of alignment tells us that every item on a page must be aligned with another item to create cohesion.

The best way to align objects is to the left or the right of the page, as the hard vertical edge makes a stronger statement. It also makes everything easier to read. When we read text, the eyes follow the text to the end of the line, and then immediately start searching for the next line. When everything is aligned, our eyes take less time to find the beginning of the next item, whether it is text or graphics.

Be careful about using centered alignment, as continuous text aligned down the center can be difficult to read, because the alignment is uneven due to different line lengths. This can tire the eyes very easily. Center alignment provides a very formal, sedate look that is not very useful for the majority of publications.

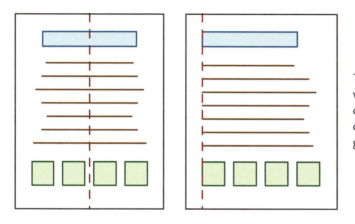

The page on the left uses center alignment, which can make long sections of text difficult to read. The page on the right creates a hard left edge of title, text and graphics.

Aligning along an axis can also be very effective, and provides a nice alternative to aligning objects along a margin. This is especially useful when you have headings placed next to text.

First heading   Aligning along an axis can also be very effective, and provides a nice alternative to aligning objects along a margin. This is especially useful when you have headings placed next to text.

Second heading   Aligning along an axis can also be very effective, and provides a nice alternative to aligning objects along a margin. This is especially useful when you have headings placed next to text.

Third heading   Aligning along an axis can also be very effective, and provides a nice alternative to aligning objects along a margin. This is especially useful when you have headings placed next to text.

Repetition and consistency

Repetition acts as a visual key to tie your publication together. Try to create a consistent and balanced look by repeating some aspect of your design throughout the entire document, such as the background, a graphic, a typeface style, a shape, or a combination of colors. This will help the reader to digest the information in your document and will help to keep their attention.

However, be careful that your repetition does not become boring. Try a little variation to mix things up and make things both interesting and comfortably familiar.

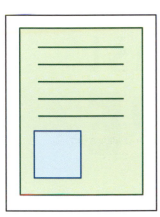
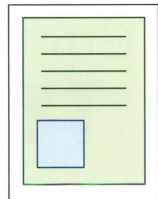
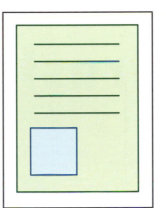

The repetition in this document looks boring and drab.

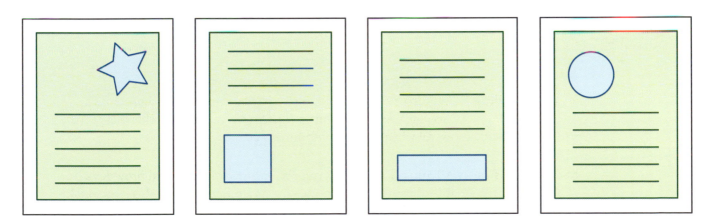

By mixing it up a little, the repetition in this document still ties the pages together, yet creates new interest.

Contrast and emphasis

Contrast and emphasis are the best ways to add visual interest to your project. This is what stands out or gets noticed first. Every layout needs a focal point to draw the reader's eye to the most important element of your document. Too many focal points will defeat the purpose. Generally, a focal point is created when one element is different from the rest, creating a visual hierarchy. The greater the difference, the greater the contrast.

You can create contrast and emphasis by using differences in size (big vs. small), type (different type styles), and color (black vs. white). Additional ways to add contrast include textures, and elements like lines, boxes, or graphics that are very different from one another. Contrast can aid in readability by making headlines and subheadings stand out.

Contrast with color

Contrast with TYPE

Contrast with SIZE

White space is a key principle of design

An important principle of design is white space. Long passages of text written edge to edge can tire the eyes. White space provides visual breathing room for the eye by breaking up text and graphics.

This is perhaps one of the hardest concepts to grasp when designing your project. Perhaps this is because white space is nothing. It is the absence of text and graphics. It is emptiness. It is negative space. But, it is not wasted space.

You can add white space to make a page less cramped, less confusing, and less overwhelming. When used appropriately, white space can improve the readability of text and make a document much easier to read. Surrounding a block of text with white space can draw a reader in, especially in a crowded layout (e.g., a newspaper) where every bit of available space is jam-packed with information.

There is no proper percentage of white space, but if a page looks or feels crowded, it probably needs more white space. Most people tend to cram too much information onto a single page, but without adequate white space, the message you want to get across will be lost.

White space does not have to be white. It can be any color. White space refers to any empty area (colored or white, opaque or transparent) that is devoid of text or graphics.

Try to achieve a balance of ink and white space using a mix of techniques described below, as appropriate to the design of your project.

Add a line of space between paragraphs

Paragraphs need to be separated, otherwise readers are faced with an endless block of text. Adding a line of white space between paragraphs helps to break the text up and makes it easier to read.

In the past, the only way to increase the space between paragraphs using a typewriter was to put two or more hard returns at the end of a line of text.¶

¶

This can be effective, depending on the effect you want to achieve. In this case, when compared with the rest of the document, there seems to be too much space between paragraphs.

These days, paragraph formatting lets you specify the amount of space to be placed before or after a paragraph.¶

In this case, the space set before the paragraph is 0.1″. This takes up much less space than a hard return, yet the space is still effective.

Indent new paragraphs

Another way to separate paragraphs is to indent the first line of the paragraph. Paragraphs following headings or subheadings do not have to be indented.

Indents can be added to the first line of new paragraphs in your page layout software.

Small indents (0.125″–0.25″) look better than indents that begin halfway across a column of text (more than 0.25″).

Do not use paragraph indents if you use spacing *between* paragraphs.

Although it is common to combine some types of paragraph spacing methods, these two together usually create too much space, or awkward spacing.

Make your gutters wider

The gutter is the inside margin or blank space between two facing pages. When the gutter is too narrow, the eye skips from a column on one page over to the column on the next page. This can make text very difficult to read. Add white space between columns by using adequate gutters.

Narrow gutters

Wide gutters

Use left-aligned text

Flush left-ragged right text alignment leaves white space at the end of each line. For more information about text justification, see Chapter 3.

Generous margins

The margins around the top, bottom, or sides of a page provide the most prominent white space in design. On a practical side, margins give the reader a place to hold the material without obscuring text or graphics with their hands. They also allow for binding the material with staples, three-ring binders, etc. Visually, margins provide a buffer zone for the text and graphics, giving the eye a break.

Add typographic white space

You can add typographic white space by:

- Increasing the leading, or line spacing, of body text. Increased leading results in more vertical space in between lines of text.
- Using a lighter type (**Arial Black is a heavy typeface** while Arial Narrow is lighter).
- Avoiding character spacing that is too tight. Tracking is the process of adjusting the space between characters, i.e., 'loosening' or 'tightening' a block of text or group of words. Adjusting tracking is useful when you want to save space, prevent widows and orphans, and improve line endings and hyphenation. However, avoid extremely loose or tight text, and avoid extreme changes in tracking within the same paragraph or in adjacent paragraphs.
- Avoiding unending condensed or heavy type.

Leave sufficient space around text wraps

Text wraps are a feature that enables you to surround a picture or a diagram with text. The text 'wraps' around the graphic. When you wrap text around graphics, or wherever text and graphics meet, provide plenty of standoff white space. Do not run text right up to the edge of graphics. Below are some simple guides for using text wraps.

Do not disrupt the flow of text

Do not let text wraps disrupt the horizontal flow of text. Readers will be forced to stop, jump over the graphic (or pull-quote), and find the correct line to continue reading.

Do not let text wraps disrupt the horizontal flow of text. Readers will be forced to stop, jump over the graphic (or pull-quote), and to continue reading. Do not let text wraps disrupt the text. Readers will be forced to stop, jump over the graphic (or pull-quote), and find the correct line to continue reading. Do not let text wraps find the correct line Do not let text horizontal flow of forced to stop, jump pull-quote), and find continue reading. Do disrupt the horizontal flow of text. Readers will be forced to stop, jump over the graphic (or pull-quote), and find the correct line to continue reading.

Visually balance space around graphics

When placing an object between two columns of left-justified text, make the standoff (the space around the graphic) slightly smaller on the left side to visually balance the space around the object.

In the top example, the ragged right alignment on the left side of the graphic leaves a lot more white space between text and graphic than on the right side where the text lines up neatly.

In the bottom example, the wrap is adjusted closer to the graphic on the left side so that the gaps from the ragged right alignment are more visually similar to the space on the other side.

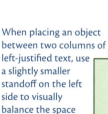

When placing an object between two columns of left-justified text, use a slightly smaller standoff on the left side to visually balance the space around the object. When placing an object between two columns of left-justified text, use a slightly smaller standoff on the left side to visually balance the space around the object. When placing an object

Ragged right alignment can create holes in the text.

When placing an object between two columns of left-justified text, use a slightly smaller standoff on the left side to visually balance the space around the object. When placing an object between two columns of left-justified text, use a slightly smaller standoff on the left side to visually balance the space around the object. When placing an object between two columns of left-justified text, use a slightly smaller standoff on the left side to visually balance the space around the object. When placing an object

Wrap space, or standoff, is adjusted to make the ragged text to the left of the object more balanced with the text on the right.

Avoid overly irregular text wraps

Do not try to follow the contours of irregularly-shaped graphics too closely. This often results in many different line lengths, too much hyphenation, and holes in the text, which can can make it very difficult to read.

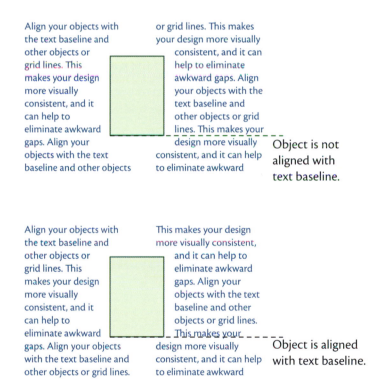

Do not follow the contours of irregularly-shaped graphics too closely. This often results in many different line lengths, too much hyphenation, and holes in the text, which can make it very difficult to read. Do not follow the contours of irregularly-shaped graphics too closely. This often results in many different line lengths, too much hyphenation, and holes in the text, which can make it very difficult to read. Do not follow the contours of irregularly-shaped graphics too closely. This often results

Text wraps can result in hyphens and large holes in text.

Align objects with baseline of adjacent text

Align your objects with the baseline of adjacent text and with other objects or grid lines. This makes your design more visually consistent, and it can help to eliminate awkward gaps.

In the top example, the bottom of the graphic does not line up with the baseline of the adjacent text.

In the bottom example, the wrap is adjusted so that the baseline of adjacent text aligns neatly with the base of the graphic. With more irregularly-shaped images, it may not be as obvious how to align the graphic and the text, but it is usually possible to find a logical and visually pleasing point of alignment.

Align your objects with the text baseline and other objects or grid lines. This makes your design more visually consistent, and it can help to eliminate awkward gaps. Align your objects with the text baseline and other objects or grid lines. This makes your design more visually consistent, and it can help to eliminate awkward gaps. Align your objects with the text baseline and other objects or grid lines. This makes your design more visually consistent, and it can help to eliminate awkward

Object is not aligned with text baseline.

Align your objects with the text baseline and other objects or grid lines. This makes your design more visually consistent, and it can help to eliminate awkward gaps. Align your objects with the text baseline and other objects or grid lines. This makes your design more visually consistent, and it can help to eliminate awkward gaps. Align your objects with the text baseline and other objects or grid lines. This makes your design more visually consistent, and it can help to eliminate awkward

Object is aligned with text baseline.

Desktop publishing

Add white space between headlines or subheads

Headlines

Headlines should be visually tied to the text below. This means you want very little space between your headline and the following text. You should have a larger space between the end of the text and the next headline.

You need less white space (leading, or line spacing) between headlines of all capital letters, than a headline that uses both upper and lower case letters. This is because there are no descenders with capital letters (i.e., the portion of some lowercase letters, such as *g* and *y*, that extend or descend below the baseline).

The space you should allow between a headline and the following text will depend on the typeface size. A general rule to follow is to make the space proportional to your headline, i.e., if your headline is 24 points, put the text 24 points beneath the headline.

Subheads

When using subheads, make sure that there is more space above than below the subhead, to tie the subhead to the text that follows it. Try using twice as much space above the subhead as below it.

Subheads should not be too close to the top or the bottom of the page. Try to have at least two or three lines of text before the subhead (if close to the top of the page), or following the subhead (if close to the bottom of the page).

If you have a subhead that wraps around to a second line, make the second line shorter than the first line. This leads the reader's eyes to the text following the subhead.

Avoid tombstoning—subheads that line up next to each other across multiple columns.

Headline spacing

You need less white space (leading, or line spacing) between headlines of all capital letters than a headline that uses both upper and lower case letters. This is because there are no descenders with capital letters (i.e., the portion of some lowercase letters, such as *g* and *y*, that extend or descend below the baseline).

The space you should allow between a headline and the following text will depend on the typeface size. A general rule to follow is to make the space proportional to your headline, i.e., if your headline is 24 points, put the text 24 points beneath the headline.

Subhead spacing

Make sure that there is more space above than below a subhead, to tie the subhead to the text that follows it. Try using twice as much space above the subhead as below it.

Subheads should not be too close to the top or the bottom of the page. Try to have at least two or three lines of text before the subhead (if close to the top of the page), or following the subhead (if close to the bottom of the page).

If you have a subhead that wraps around to a second line, make the second line shorter than the first line. This leads the reader's eyes to the text following the subhead.

HEADLINE SPACING

You need less white space (leading, or line spacing) between headlines of all capital letters than a headline that uses both upper and lower case letters. This is because there are no descenders with capital letters (i.e., the portion of some lowercase letters, such as *g* and *y*, that extend or descend below the baseline).

The space you should allow between a headline and the following text will depend on the typeface size. A general rule to follow is to make the space proportional to your headline, i.e., if your headline is 24 points, put the text 24 points beneath the headline.

Subhead spacing

Make sure that there is more space above than below a subhead, to tie the subhead to the text that follows it. Try using twice as much space above the subhead as below it.

Subheads should not be too close to the top or the bottom of the page. Try to have at least two or three lines of text before the subhead (if close to the top of the page), or following the subhead (if close to the bottom of the page).

If you have a subhead that wraps around to a second line, make the second line shorter than the first line. This leads the reader's eyes to the text following the subhead.

OTHER DESKTOP PUBLISHING AND DESIGN TIPS

Use rough thumbnail sketches to explore layout options

Thumbnail sketches are rough drawings used to explore layout options. These quick pen or pencil sketches let you try out several ideas and determine the most likely layouts before beginning a project. This step can sometimes save you a lot of time and effort spent rearranging text and figures in a design program. The thumbnails below show a possible layout for a four-page newsletter.

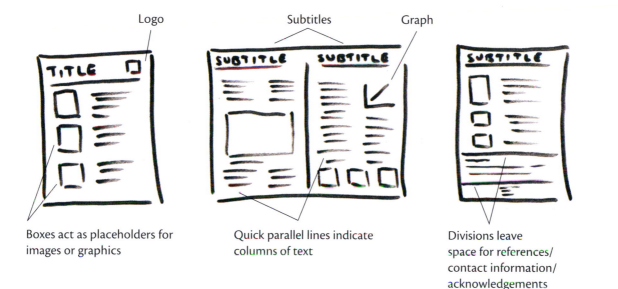

Logo

Subtitles   Graph

Boxes act as placeholders for images or graphics

Quick parallel lines indicate columns of text

Divisions leave space for references/ contact information/ acknowledgements

Designing with thumbnail sketches

- Do not go into too much detail. Use thumbnails to establish approximate locations for major elements.
- Try for an approximately proportional page size. You do not have to get out the ruler. This is meant to be general.
- Make lots of rough sketches. You will rule out many design ideas quickly this way before wasting time in a page layout program.
- Do not try to do these initial rough designs in your software, even if using dummy text and placeholder graphics. You will be apt to get caught up in little fiddly things like changing the typefaces or aligning graphics perfectly. You can save that step for after you have finished the initial brainstorming for ideas, using thumbnail sketches.

Hints for text-rich pages

Very dense pages of text can be very hard to read, and will intimidate a potential reader. There are several ways to rescue text-rich pages.

- Break your page into three columns, and leave one column as white space to counterbalance the text.
- Add a drop cap, where the first letter of a paragraph takes up the equivalent of two or three lines of text.
- Use headers and footers to let the reader know where they are in a document.
- Add subheads to help organize your text and break it up into digestible chunks.
- Pull-quotes can help to emphasize a key point of interest.
- Boxes and bulleted lists can help to break things up.
- Add graphics, photos, or diagrams that help to illustrate your key message.

Whatever you decide to use, make sure that your text organizers are consistent.

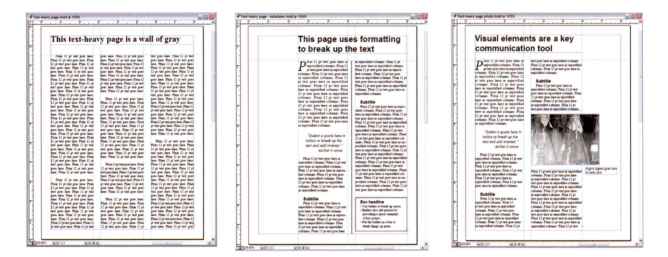

The text-heavy page on the left is a wall of impenetrable gray. It is very easy to break this up using a variety of the tools listed above. The best way to add interest to your project is to incorporate graphic elements.

Graphics are not fillers—they are a key way to get your message across

Research shows that people remember the graphic elements of a publication before they remember anything else. There are a few simple tips to make your graphics effective.

- Make sure you use high quality graphics with a minimum resolution of 300 DPI for printing. Low quality images can be blurry and distracting.
- Scan your images using a resolution of 300 DPI at the final dimensions at which you will be using them.
- For professional printing, convert all RGB images to CMYK.
- Do not have action going off the page (e.g., someone in a photo walking *off* the page).
- Text superimposed over images can be hard to read—an easy trick to fix this is to lighten up the photo and use a dark typeface.

Create a headline hierarchy

Using headlines and subheads in a document is an effective way to organize information. It also helps the reader to approach your document by breaking up text into easy-to-digest chunks. The size of headlines and subheads helps to establish importance in a document. If you are using multiple headlines and subheads, create an obvious hierarchy using typeface, size, and color. You need to make sure that there is a distinct difference between each headline or subhead level.

MAIN HEADLINE

Subhead 1

Subhead 2

Text

Bleeds

Bleeds are used when you want an element of your document, such as a colored bar, to extend right up to the edge of the page, without a white margin around it. The proper way to do bleeds is to make that element extend off the page by around 0.25″. The printer will print your document on larger-sized paper, then trim it using a guillotine to the right size, resulting in the colored bar extending right up to the edge of the page. The reason you need 0.25″ extra is because the trimming process is not exact, and bleeds leave a margin for error. In desktop publishing software, such as Adobe InDesign, you can specify how much bleed you want in each direction (usually 0.25″ top, bottom, inside, and outside) and then you will see the bleed guides on your page spreads.

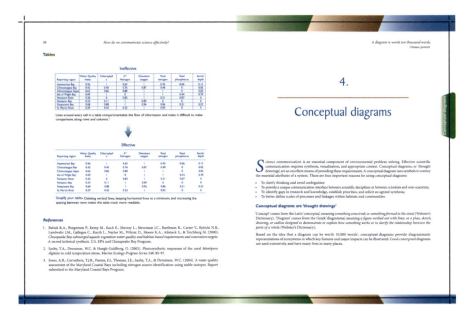

This is a double-page spread set up using bleeds of 0.25″ in each direction. The black box represents the final trim size of the document (in this case it is tabloid size—11″ x 17″). You can see the colored bar extending outside of the black box to ensure that the color will extend to the edge of the page once the page is printed and trimmed, like the example below.

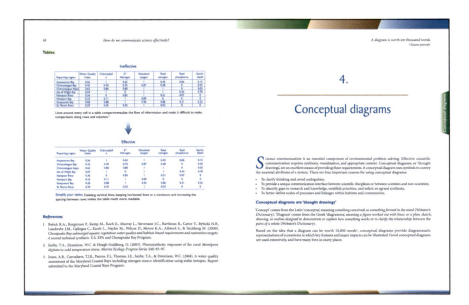

Consider your column width before choosing typeface size

Lines of type that are too long or too short can slow down reading and comprehension. Type size, margins, gutters, and the number of columns help to determine line length in your publication.

Large type in narrow columns can be difficult to read as more words end up being hyphenated. The eye also has to keep trying to follow the flow of text, which can make reading difficult. Try to keep line lengths between 45 and 75 characters.

For narrower columns you should use smaller type. For narrower columns you should use smaller type. For narrower columns you should use smaller type. For narrower columns you should use smaller type. For narrower columns you should use smaller type. For narrower columns you should use smaller type. For narrower columns you should use smaller type. For narrower columns you should use smaller type.

For wider columns, you can use larger type. For wider columns, you can use larger type. For wider columns, you can use larger type. For wider columns, you can use larger type. For wider columns, you can use larger type. For wider columns, you can use larger type. For wider columns, you can use larger type. For wider columns, you can use larger type. For wider columns, you can use larger type.

Text with borders needs space

When using borders or putting text inside boxes, do not allow the text to get too close to the border. Instead, give the text margins inside the box.

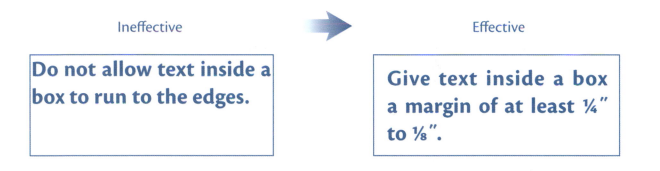

Ineffective

Do not allow text inside a box to run to the edges.

Effective

Give text inside a box a margin of at least ¼" to ⅛".

Do not colorize text smaller than 12 point

All printing presses have a little variation in consistency of position of the different color plates. This is called misregistration, i.e., the cyan, magenta, yellow, and black do not line up exactly, so you end up with little colored haloes around the characters.

Colorizing small text can result in misregistration.

FURTHER INFORMATION

The books and websites below are current at the time of printing. However, as software is constantly upgraded and changed, these books and websites may also change and be updated. This is not an exhaustive list, but a small sample of useful resources.

Adobe Creative Team. 2006. *Adobe InDesign CS2 classroom in a book.* Adobe Press, Berkeley, California, U.S.A.

Blatner D, & Cohen S. 2005. *Adobe InDesign CS2 tips and tricks.* Adobe Press, Berkeley, California, U.S.A.

Kelby S, & White T. 2005. *InDesign CS2 killer tips.* Peachpit Press, Berkeley, California, U.S.A.

Parker RC. 2003. *Looking good in print (5th ed.).* Paraglyph Press, Berkeley, California, U.S.A.

White AW. 2002. *The Elements of Graphic Design: Space, unity, page architecture, and type.* Allworth Press, New York, New York, U.S.A.

Integration and Application Network   *www.ian.umces.edu*

Adobe Systems Incorporated   *www.adobe.com*

Designer Today Graphic Design Magazine   *www.designertoday.com*

My Design Primer   *www.mydesignprimer.com*

About Desktop Publishing   *www.desktoppub.about.com*

DesktopPublishing.com   *www.desktoppublishing.com*

Tell them what you're going to tell them. Tell them. Then tell them what you told them.

Anonymous

6.

Posters and newsletters

Posters and newsletters are common ways to communicate science to a broader audience. Similar principles can be applied to both posters and newsletters, as both should be graphic-rich, with a judicious amount of text that supports the graphics, not the other way around.

POSTERS

Posters are a common form of science communication, especially scientific posters at conferences. A poster session at a large conference might have as many as 500 posters. Making your poster stand out and attract attention among so many others is the key to a successful scientific poster.

Choosing the right software

The first choice for software to produce posters is desktop publishing software, such as Adobe InDesign or Quark XPress. Dedicated desktop publishing software allows flexibility of layout, and allows graphics files to be linked, without directly embedding them. This means that total file size is smaller, making it faster to work with on your computer. Ensure that linked image and graphic files are in the same folder as the poster file, and always supply the printer with a high resolution PDF just in case there are problems with the links. For precision graphics, create your images in vector-drawing software, such as Adobe Illustrator or CorelDRAW, and link the vector EPS files into your desktop publishing program. Vector drawing software, such as Adobe Illustrator or CorelDRAW, is the next best choice to create posters, with powerful graphic options. However, these programs are not as powerful as true desktop publishing software when it comes to handling text and linked graphic files.

Microsoft PowerPoint should be avoided to make posters. Because it is designed for on-screen presentations, the files are in RGB format, not CMYK, which causes problems when printing the poster. Printing ink-based color (CMYK) from light-based color (RGB) inevitably results in the printed result being vastly different to what you see on screen (see Chapter 3 for more information on RGB and CMYK color formats). Graphic files cannot be linked—they have to be embedded within the PowerPoint file. This dramatically increases file size and makes the file slow to manipulate on your computer. Graphics are not handled as powerfully as with dedicated desktop publishing software, and can be hard to align and size properly, and text flow lacks kerning, tracking, and other spacing controls found in desktop publishing software. Using different versions of PowerPoint can mean that text and graphics in the poster may move if you send the file to a different computer, depending on the preferences of that computer. Also, the maximum dimensions of a file in PowerPoint is 56 inches, which may be restrictive depending on your needs.

The title is your key message

Depth range and light requirements of the seagrass *Zostera capricorni* in Moreton Bay, Australia.

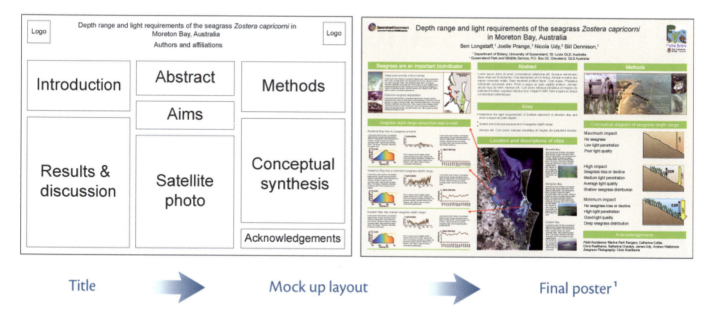

Title ➡ Mock up layout ➡ Final poster[1]

- It is best to begin with the fundamental message—this will be the title.
- Assemble the visual elements that will explain and support the main message.
- Add a small amount of text to support the graphics.
- Draw 'mock up' layouts to help plan flows as well as the balance between images and text.

Layout depends on your audience

Posters are possibly the most widely accessed form of science communication, as dozens or even hundreds of people can view one well-placed poster. This opens the possibility of communicating to a broad audience, and different audiences require a different balance of visual elements to facilitate communication. For the general public, attractive but relevant and informative photographs are essential and should capture the main message. For the informed public or non-specialist scientists, conceptual diagrams and maps provide more detailed information, but are widely understood. A methods paper to a technical scientific audience may require more text, but visual elements should still capture the essential message.

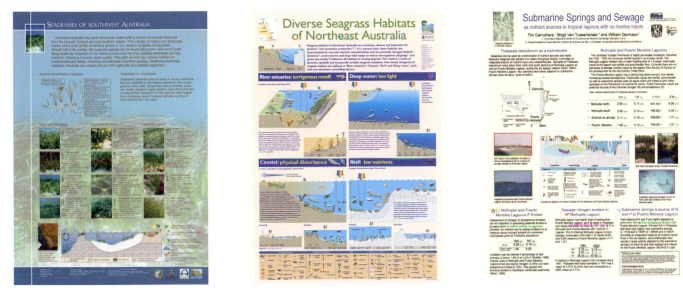

General public[2]

Informed public or
non-specialist scientists[3]

Technical specialist scientists[4]

If in doubt, use a plain background. Photographs can sometimes be used effectively as a background to a poster, however, often they only distract the audience and add nothing to the message of the poster. Exceptions are posters designed to communicate to the general public, where a relevant and attractive photograph can be an excellent way to capture the attention of a passer-by. For more informed or scientific audiences, however, the combination of other visual elements within the poster should be designed to be sufficiently captivating.

Posters & newsletters

Titles and authors should be large and clear

Using active voice when writing poster titles implies confidence in results and will make the reader want to know more. An example of an active title that makes a statement is *Salt-excreting species are the most susceptible*, rather than the boring title *Results*. Another way to create an effective title is to use a short question as the title, such as *Do stream herbicides affect mangrove photosynthesis?* The reader will presume it is rhetorical and will be drawn to the poster to obtain the answer. Include the authors' first and last names, and institutions (institution and country are usually sufficient).

Do not hide the author list down at the bottom of the poster in a tiny typeface—people will assume that you do not want to be associated with the work. Institutional logos should be inclusive, clear, and of high quality—avoid grabbing these from the internet at the last minute, as these will be low resolution files.

Ineffective

Effective

The title is the single most important part of a poster, and in this poster, the title is far too small and hard to read. This title is also too long and gives no clue as to the conclusions. It is difficult to read the authors' names, the background photo here is distracting, and there are no figure legends or photo captions.

The title must be large enough to read from 30´ (10 m) away, which means the letters should be between 72–96 point font size. The title should be as short as possible, be active (preview your conclusions), and state your main message. Subheadings should also be active statements. Authors and affiliations are the second most important piece of information on a poster. Authors' names should be easily legible. A photo of the principal can be useful in identification. A plain colored background and self-contained photo captions enhance readability. [5]

Clear, large text should support visual elements, not the other way around

Self-contained visual elements within the poster should tell the message and limited text should be used to support these graphics. Text should be written in a clear, consistent, system typeface (e.g., Times or Arial) and be 18–24 point size. The smallest typeface size you should use on a poster is 12 point, and that should only be for text such as the reference list and image credits.

Ineffective

Effective

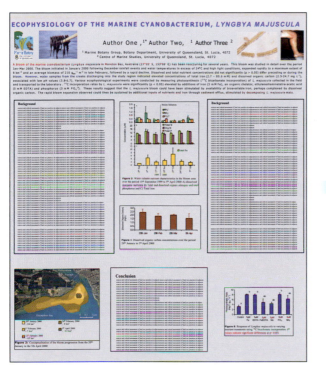

Too much text printed in a small font will not be read, and reduces the information that your audience takes away with them. Photographs and images should at least have a caption to link them to the rest of the poster—do not add decoration as this will distract the reader from the main message.

As posters are often viewed without the benefit of the author being present, it is important to provide context for the work. An excellent way to do this is to provide good maps or satellite photos as well as photos of the organism. Graphics should provide the main focus of the presented message, and should only be supported by text—preferably as dot points or short statements.[6]

Use active titles

Unlike journal papers, posters have no editors, so there are no hard and fast rules on format, content, and style (notwithstanding the advice and hints already given). The benefit is the potential for flexibility and originality in finding the best way to capture an audience and communicate a message. An effective technique that can be used is to replace the staid and largely uninformative *Methods, Results,* and *Discussion* with active titles and subtitles that contain the main point that is being communicated. These subheadings should provide the essential statements that support the overall conclusion (in the title). If a reader is still interested, they will read the rest of the text or ask a question. This approach also helps to clarify your own thinking, as each paragraph needs to be synthesized, and so superfluous information quickly becomes apparent.

Use two or three main, subtle colors

Color is a very useful tool, but should be applied carefully. A good contrast between a neutral background and one or two main text colours is appropriate. Brighter colors can be used to create links between data in tables, site maps, photographs, and main conclusions. Symbols can also be used to create linkages and improve clarity. Conceptual diagrams are also well suited to posters as they capture main messages and include self-contained legends. They also provide a compelling presentation of the overall message and clearly show linkages between different elements of the systems being discussed. All of these techniques help the reader to navigate the poster, ensuring that they observe the most important points, and become familiar enough with the issues, to question further.

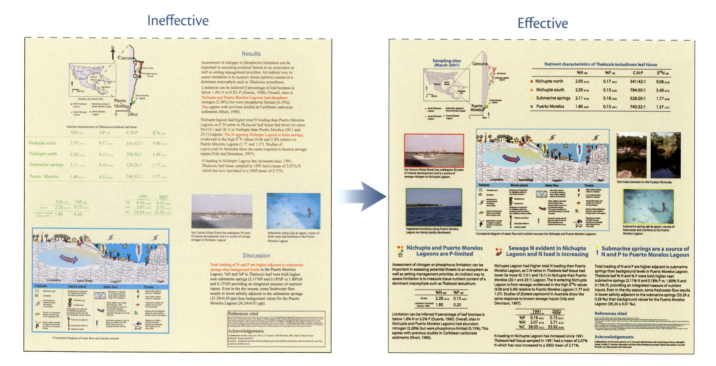

An ineffective poster with bad positioning, inappropriate use of color, and passive, non-informative subheadings.

An effective poster with color used to connect text and graphics. In this case, green, red, orange, and yellow are used to connect the site locations on the map (top left) with the results table (top right), with the photos (center), and with the conclusions (bottom). Active, informative subheadings, and subtle background color scheme make navigation of the poster easier. [4]

Do you have time to make a poster?

Producing a good poster takes time. The effort is similar to producing a presentation from scratch, and the benefit is a tangible end product which can be displayed or re-used. Once all data is analyzed, a standard data poster will take four to five days to get print-ready. Larger posters may take up to 10 days or even longer, if all the graphics have to be created. The time, however, is worth investing as a bad poster is a waste of your time and will not communicate the main messages effectively.

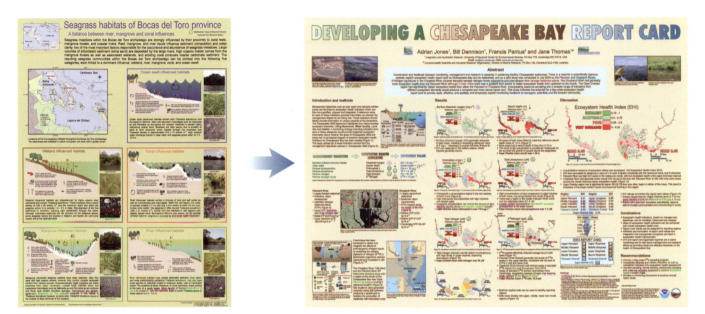

Four to five days to produce. [7]   Eight to 10 days to produce. [8]

Handouts will increase the longevity of a poster's message

Providing handouts at a meeting is an excellent way of ensuring that the message of the poster lives on past the meeting, as people are often overloaded with too much information from the many posters and presentations they have attended. Handouts also provide contact details for follow up in the future. One simple and effective way to produce a handout is to print small (8.5″ x 11″, or A4 size) versions of the poster—in either color (preferably), or black and white. A well-designed poster should still be readable at this size. Ensure that at least names, addresses, websites, and most importantly, email addresses are clearly legible on the handout. If time permits, a separate handout can be made with the poster title, abstract, and one or two key graphics, as well as addresses and affiliations.

The bottom line—essential rules and tips for making posters

- The title needs to be concise. It should contain the key message, and be large enough to read from 10 m away.
- Authors and affiliations need to be central and clearly legible.
- Use self-explanatory, stand-alone graphics where possible.
- A poster is an advertisement, not a review paper—use text sparingly and only to support graphics.
- Use color judiciously—two or three colors is best (use additional colors only for highlights).
- Be aware of the audience, in assessing how best to capture their attention (balance of photos, graphics, and text).
- Think twice before using a photo as a background—it is usually just a distraction.
- Take the time to review the layout, ensuring the whole poster supports the title.
- Provide handouts of the poster.
- Create the poster file using desktop publishing software at the print size (e.g., 3′ x 4′), and always use CMYK color format for printing (see Chapter 3 for more information on CMYK).
- Provide the printer with a high resolution PDF of the poster, as well as the desktop publishing poster file, and any graphic files that are linked into your poster.
- Either create outlines of text or ensure you provide the printer with all typefaces used in text, or do not use non-system typefaces.

NEWSLETTERS

Newsletters are a very effective way of communicating to a broad audience. A full-color, four-page newsletter can convey a wealth of information and can include a combination of visual elements—conceptual diagrams, satellite photos, maps, photographs, and data figures.

Many of the principles for creating posters also apply to newsletters, both being desktop publishing products. Professional desktop publishing software should be used where possible. A hand-drawn mock-up layout will help to determine the locations of the main elements.

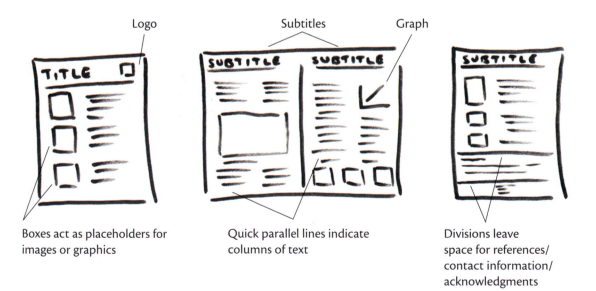

Boxes act as placeholders for images or graphics

Quick parallel lines indicate columns of text

Divisions leave space for references/ contact information/ acknowledgments

Newsletters are generally designed to be accessed by a wide audience. The multiple pages allow for explanation of complex or difficult concepts, with the aid of visual elements, and a reference list or list of further readings is appropriate. The titles and subtitles should be active statements, and be clearly visible. The front page should contain the abstract, or a summary of the newsletter's message. The bottom of the back page is a good place to put a reference list, contact information, and acknowledgements.

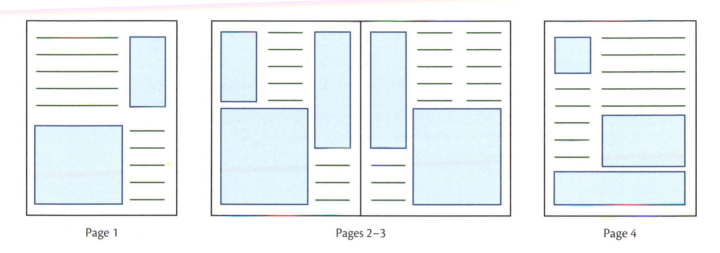

Page 1   Pages 2–3   Page 4

An example of a four-page newsletter, laid out using a general base of three columns. Text boxes and visual elements can take up one, two, or three columns' width without compromising the overall three column layout.

These are two examples of science newsletters. These newsletters utilize different layouts, depending on content, and employ a variety of visual elements to ensure a clear message.[9,10]

Posters & newsletters

REFERENCES

1. Longstaff BJ, Prange J, Udy N, Drew E, & Dennison WC. 2000. *Depth range and light requirements of the seagrass* Zostera capricorni *in Moreton Bay, Australia.* Poster presentation at the International Symposium on Ecosystem Health, Brisbane, Australia.
2. Cambridge M, Carruthers TJB, Saxby TA, Dennison WC, & Robb M. 2005. *Seagrasses of southwest Australia.* Poster prepared for Coastcare Western Australia.
3. Carruthers TJB, Dennison WC, Longstaff BJ, Waycott M, Abal EG, McKenzie LJ, & Long WJL. 2001. *Seagrass habitats of north eastern Australia: models of key processes and controls.* Poster presentation at the Australian Coral Reef Society conference, Magnetic Island, Australia.
4. Carruthers TJB, van Tussenbroek B, & Dennison WC. 2002. *Submarine springs and sewage as nutrient sources to tropical lagoons with no riverine input.* Poster presentation at the International Seagrass Biology Workshop v, Ensenada, Mexico.
5. Bell AM, & Duke NC. 2002. *Effects of photosystem II-inhibiting herbicides on mangroves.* Poster presentation at the Canopy Conference, Cairns, Australia.
6. Watkinson AJ, Dennison WC, & O'Neil JM. 2000. *Ecophysiology of the marine cyanobacterium* Lyngbya majuscula. Poster presentation at International Society for Ecosystem Health Meeting, Brisbane, Australia.
7. Carruthers TJB, Jacome GE, & Barnes PAG. 2003. *Seagrass habitats of Bocas del Toro province.* Poster presentation at the 31st Scientific Meeting of the Association of Marine Laboratories of the Caribbean, Trinidad.
8. Jones AB, Dennison WC, Pantus F, & Thomas JE. 2004. *Developing a Chesapeake Bay report card.* Poster presented at the National Council for Science and the Environment's 4th National Conference on Science, Policy and the Environment, Washington, D.C., U.S.A.
9. Chesapeake Bay Program and the Integration and Application Network. 2005. *Chesapeake Bay environmental models.* Integration and Application Network newsletter #11, Maryland, U.S.A.
10. Submerged Aquatic Vegetation workgroup and the Integration and Application Network. 2005. *Bay grass restoration in Chesapeake Bay.* Integration and Application Network newsletter #12, Maryland, U.S.A.

FURTHER INFORMATION

The websites below are current at the time of printing. However, websites constantly change and are updated. This is not an exhaustive list, but a small sample of useful resources.

Scientifically Speaking: Tips for preparing and delivering scientific talks and using visual aids
  http://www.tos.org/resources/publications/sci_speaking.html
Advice on designing scientific posters   *www.swarthmore.edu/NatSci/cpurrin1/posteradvice.htm*
Instructional graphic guidelines   *www.uwstout.edu/lts/graphics/guidelines.shtml*
How to make a great poster   *www.aspb.org/education/poster.cfm*

7.

Presentations

Giving presentations is a fact of life for scientists. Whether you are giving a formal presentation to a large conference, or a more informal talk to your local community, presentations provide an opportunity to personally present your data to an audience. Remember that presentations are supposed to facilitate interaction with your audience as a means of two-way communication, so be prepared for, and encourage questions and comments. People often find giving presentations stressful, but proper preparation and practice will help make you as ready as you can be.

COMMUNICATING YOUR MESSAGE EFFECTIVELY

Arrive early to the venue where you will be giving your talk. It is not polite to make your audience wait for you—you should plan to be the first to arrive. This allows you to assess the layout of the room, where the computer and projection screen are, and how the audience will be arranged. Load up your talk on the computer you will be using as soon as possible, to check that all the equipment is in working order. Bring as much back-up equipment as possible, including an extra copy of your presentation (on CD or a USB memory stick), and if possible, your own laptop computer, data projector, and power cords and extensions.

When working on your talk, generally allow *one minute per slide* on average, and allow time for questions. If you have one hour alloted for your presentation, plan on 45–50 slides in 45–50 minutes, and allow 10–15 minutes at the end for questions and discussion. Your audience will start getting restless if you go over your alloted time, and it will limit the interactive discussion time after your talk. Most conferences have strict timelines to adhere to, and will cut you off if you go over your alloted time, whether you have presented all your information or not. Make sure you are aware of how much time you have alloted to you, and stick to it.

Preparation is the key. Do not memorize a speech, but let well-prepared slides provide the cues that you need. The text on your slides should function to remind you of the key points of your talk and provide the audience with enough information to interpret the graphics, but should not replace the need for you to talk or distract the audience from what you are saying. Remember that your audience will usually read a new slide before they start listening to you explain it.

Provide appropriate background for your audience. It is better to err on the side of not assuming too much prior knowledge, but always treat your audience with respect and never insult their intelligence. A good rule of thumb is to assume naïveté, but not ignorance. This is especially true when giving presentations to people who are not scientists—never underestimate people's ability to understand your story just because they are not professionals in your particular field. It is your responsibility to be able to communicate your work to a broad audience. It is not your audience's responsibility to try to wade through overly technical explanations, raw data, and bad graphics. Instead of 'dumbing it down', your communication should be of a high enough standard to inform and educate people.

Show data within five minutes, no matter how long the presentation. Scientific presentations are all about your data. You need to introduce the audience to your topic, but do it in five slides or less and then move on to the really interesting stuff—your data! Continue giving the audience one good idea at least every 10–15 minutes. Any longer than that between new concepts and your audience will start getting bored.

Orient the audience to each slide, by explaining the axes and units on each graph or figure, and what each visual element (photos, conceptual diagrams, etc.) represents, and why it is relevant. Usually by the time you give a presentation, you have looked at it so many times that it is very familiar to you. Do not forget that it is not familiar to your audience.

Tell them what you are going to tell them. Tell them. Then tell them what you told them. Summarize as you progress through your talk, building on information and data that you have already presented to tell your story. Use active slide titles to provide ongoing summaries, e.g., instead of using *Seagrass light requirements*, or the even more boring *Results* as slide titles, a title like *Seagrasses need 10% of incident light* is an active statement and sets the scene for the slide. Regular reinforcement of the concepts and results is a good idea as you build your story throughout your presentation, especially as your audience cannot refer back to earlier parts of your talk unless you give them a handout.

A rule of thumb is that almost *every slide should have visual elements*, not just dot points. Presentations are a very visual medium. Your audience will quickly become bored when presented with slide after slide of text. Figures, photos, and conceptual diagrams contain much more information than just text, and they provide the audience with a visually interesting backdrop while you explain their relevance. Make sure all your visual elements are well designed and formatted to be visible on the screen. How many scientific presentations have you seen where the presenter shows a graph or other figure that is very difficult to see, and then says, "This figure didn't really come out very well" or "You can't really see this graph, but I'll tell you what's on it"? With proper attention to your graphics, there is no reason why everything on your slides should not be perfectly visible to your audience. If you cannot format your graphic so that it is visible, leave it out altogether.

Practice in front of colleagues—once for feedback on the structure of the presentation and layout of the slides (ask somebody to take notes for you so you can edit your slides later), and again to rehearse timings.

Most people are *nervous* before and during presentations. Apart from sufficient preparation, there are a couple of things you can do to make the presentation go smoothly. If you are using a laser pointer, make sure to hold it in both hands when you are using it—it will keep the beam steady if your hands are shaking. Also make sure to have a glass of water nearby during your presentation, to relieve a dry mouth.

COMPONENTS OF A GOOD SCIENTIFIC PRESENTATION

The title slide contains the most important parts of a scientific presentation: the title and the author/s and their affiliations and logos. The title slide also serves as an introduction to you and your research. Your title slide will often be loaded up on screen while your audience is arriving, so having a relevant photo as the background can be appropriate.

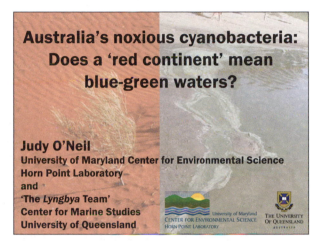

This is an example of an effective title slide. The photos in the background provide visual interest without being overpowering (this is done by adding a semi-transparent white box over the photos), and are tied to the title of the presentation. The authors and their affiliations are indicated by the logos.

An outline slide is a guide to your talk. It explains the organization of the presentation and can preview the conclusions—but do not give away your whole story at the beginning! It is often useful to prepare your outline slide first of all—you can figure out the structure of your presentation, and then prepare the rest of the slides.

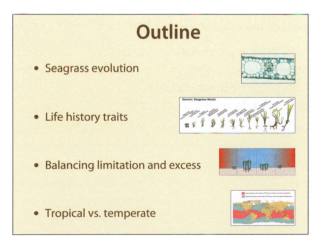

An effective outline slide gives the audience a preview of your talk. You can include thumbnail images of key graphics, or even thumbnails of individual slides.

The body of the presentation includes an introduction to the topic. Your introduction slides explain how your study fits in with the current knowledge, outlines your hypotheses, and states the objectives of your study. Photos of the organism or habitat that you are studying introduce your audience to your research. Slides of your methods can include visual elements. Combinations of diagrams, photos, maps, and satellite images will orient the audience to your study site and experimental approach. Results slides can be data-rich with graphs and tables providing both detail and the big picture. Data summary slides can use conceptual diagrams to synthesize your data. Summarize as you go—as you show more results, put them in the context of your other results, and the results of other studies.

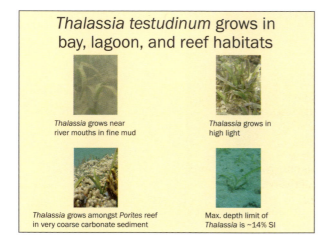

An effective introduction slide, with photos of the organism in question (the seagrass *Thalassia testudinum*).

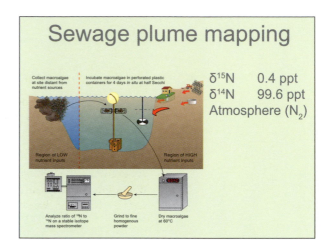

An example of an effective methods slide. A conceptual diagram summarizes the procedure used to obtain and analyze the samples.

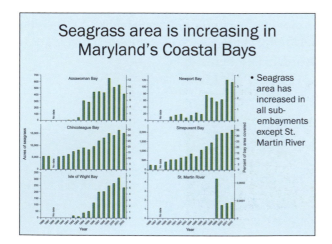

An effective results slide, with the six graphs supporting the statement made by the dot point.

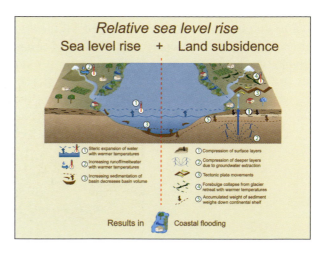

An example of an effective data synthesis slide. A conceptual diagram is a good way to summarize the salient points of the presentation.

Your conclusions are the take-home messages. A series of dot points may be accompanied by a relevant background photo and should stimulate questions from the audience.

An example of an effective conclusions slide. The series of dot points sums up the contents of the presentation, with a relevant background photo.

Use an acknowledgements slide to list study participants by category or by institution. Specify the roles of the different people and organizations. A background photo can be appropriate here, as this is usually the last slide of the presentation and so may be up on screen during the following discussion time.

An effective acknowledgements slide lists people and institutions. A background photo may be appropriate.

Prepare extra slides and include them at the end of your talk. Try to anticipate what questions your audience may ask, and prepare these extra slides with more information or data that might address those questions. If you do get asked the questions, you can quickly load up the extra slides to illustrate your answer. Answer any questions to the best of your ability, but do not be afraid to say, "I don't know". You can, however, follow this with any information that you *do* know that is relevant to the question.

FORMATTING YOUR SLIDES EFFECTIVELY

Consistency of slide background, typefaces, size, and layout between slides is important. Consistency will help your audience follow your story by preventing them having to reorient themselves to each slide. It is a good idea to use the *Master Slide* function in your presentation software to specify the formatting and layout to be applied to all the slides. Every time you create a new slide, the Master Slide formatting will automatically activate.

Typeface size should be at least 18 point for bullet points to be visible. The typeface size can be smaller than this (down to 12 point) for small text, such as citations and photo credits. Slide titles need to be big—at least 36 point. As presentations are viewed from a distance and generally contain relatively little text (compared to a manuscript, for example), using sans serif typefaces is a good idea (see Chapter 3 for more information on typefaces).

Elements need to be aligned with each other. Text boxes, photos, graphs, tables, and graphics should be aligned both horizontally and vertically. Symmetry is important as well—try to make your elements the same size as each other where possible. Photos should be cropped to best emphasize the point of the photo—unnecessary parts of a photo are distracting.

Ineffective | Effective

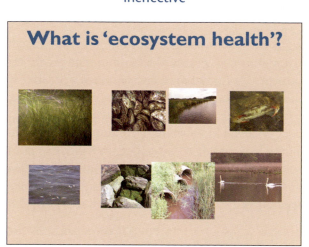

The slide title gives a clue about what the photos are supposed to be representing, but no bullet points or details give any information about the photos. They are also not cropped, resized, aligned, and distributed properly.

A few well-chosen bullet points answer the question in the title. Subheadings above each row of photos make the point of the photos clear. The photos have all been cropped, resized, aligned, and evenly distributed both horizontally and vertically.

Presentations commonly use bulleted lists to convey information. If you have text that you want in a bulleted list, ensure that the text is aligned properly on the left as a 'hanging' indent.

- This text is not formatted as a proper 'hanging' indent. Subsequent lines of text align with the bullet on the left, instead of aligning with the start of the first line of text.

- This text is properly formatted as a 'hanging' indent. All the lines of text align on the left, with the bullet 'hanging' outside the text.

Graphs should be formatted before inserting them into your presentation software, by removing all background color, gridlines, and borders, and using complementary colors. For more information on effective graph formatting, see Chapter 3. For more advanced editing of your graphs, you can copy and paste the graph into a vector-drawing program such as Adobe Illustrator or CorelDRAW. You can then add gradients and shading of colors, additional objects such as photographs and conceptual diagrams, and other graphic features.

Ineffective

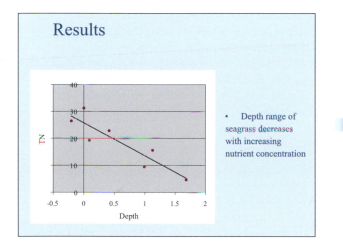

Effective

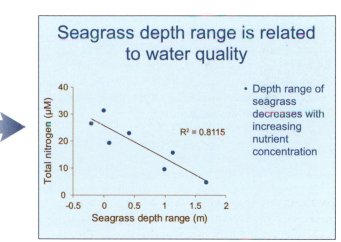

The passive slide title 'Results' gives no clue as to the contents of the slide. The serif typeface used (Times New Roman) is harder to read at distances. The lack of alignment and symmetry within the slide looks messy, and the text box does not have a 'hanging' bullet. The formatting of the graph obscures the data—the gridlines, and gray and white boxes add clutter but do not increase the data density.

An active slide title, 'Seagrass depth range is related to water quality', provides the point of the slide. Changing the typeface to the sans serif Arial improves readability. Centering the slide title, and aligning the top of the graph with the top of the text box provides alignment and symmetry. The text in the text box is properly aligned on the left-hand side, with the bullet 'hanging' outside the text. Careful data formatting makes the message clear— the graph is transparent so the slide background is visible, the axes are well labeled, and the colors match the rest of the slide.

Dark-colored text on top of a light background is much easier to look at for long periods than light text on a dark background. Some examples of effective and ineffective slides follow. For more information on color combinations, see Chapter 3.

Ineffective

Effective

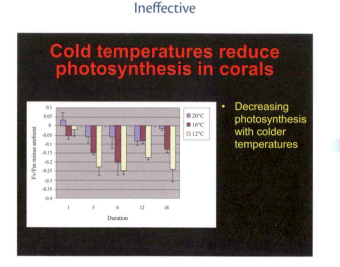

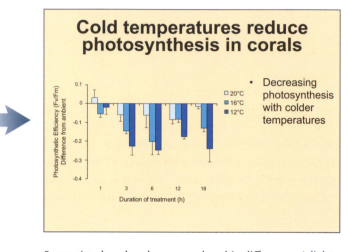

The color scheme of this slide (black background with red and yellow text) clashes badly and would be difficult to look at for long periods. The graph has not been formatted correctly—the gridlines, and white and gray boxes add clutter but do not increase the information density. The axes are not labeled adequately with the measurements they are depicting, or the units in which the measurements were made.

Some simple color changes make a big difference. A light background (pale yellow) and dark text (black) provide effective contrast while not being tiring to the eyes. Careful data formatting makes the message clear—the graph is transparent so the slide background is visible, the axes are well labeled, and even the colors in the graph (blue) are relevant to the experimental conditions (cold temperatures).

Animation functions in your presentation software can be useful, for example, to show steps in your methodology, or a sequence of maps over time. However, avoid the built-in special effects such as animations that zoom in from the side or with sound effects. These types of special effects are inappropriate for scientific presentations and erode your credibility with your audience. Let your data tell its own story. The focus should be on your content, not on the tool you are using to present your content.

One of the most common problems with presentations is that the file size of your presentation can quickly become very large, especially if a large number of photos are used. This can cause problems in transferring files between computers, and slow down loading time and slide changes during your presentation. Graphics and images do not need to be print-quality resolution to be perfectly visible on the screen. Your images need to be 96 DPI (dots per inch) at whatever size you will be using them on-screen. The best way to minimize file size is to reduce the resolution of your graphics *before* inserting them into your presentation.

Keep in mind that at conferences and workshops, you will often have to present your talk using a different computer to the one you used to create your presentation. This is a good reason to only use standard system typefaces in your presentation. If you have used any non-system typefaces, it is likely that these typefaces will not be installed on the computer you will be using to give your presentation. If this happens, the computer will automatically assign system typefaces to replace the typefaces that are not installed, which usually messes up your slide layout.

Movie and video clips are very effective visual elements to use in presentations, and you can easily insert short video clips into your presentations. Presentation software does not usually embed movie files into presentations, so you will need to take a copy of the movie file along with your presentation file, otherwise your presentation will not be able to find the movie file. This is another good reason to rehearse your presentation on the computer you will be using to give your presentation.

FURTHER INFORMATION

The books and websites below are current at the time of printing. However, books and websites constantly change and are updated. This is not an exhaustive list, but a small sample of useful resources.

Tufte ER. 2003. *The cognitive style of PowerPoint.* Graphics Press, Cheshire, Connecticut, U.S.A.

Scientifically Speaking: Tips for preparing and delivering scientific talks and using visual aids
http://www.tos.org/resources/publications/sci_speaking.html

The Gettysburg PowerPoint Presentation   *www.norvig.com/Gettysburg*

Presentation Pointers   *www.presentation-pointers.com*

About Desktop Publishing   *www.desktoppub.about.com*

Presentation Helper   *www.presentationhelper.co.uk*

Hints for giving seminars with PowerPoint   *www.swarthmore.edu/NatSci/cpurrin1/powerpointadvice.htm*

Instructional graphic guidelines   *www.uwstout.edu/lts/graphics/guidelines.shtml*

Creating better presentations   *www.easternct.edu/smithlibrary/library1/presentations.htm*

You can have brilliant ideas, but if you can't get them across, your ideas won't get you anywhere.

Lee Iacocca

8.

Websites

Web design is the art and process of creating a single web page or an entire website and involves both the aesthetics and the mechanics of a website's operation. Some of the aspects of web design or web production include graphics and animation creation, color selection, typeface selection, navigation design, content creation, coding of the HTML (Hypertext Markup Language) and various other scripting languages, and database development. Web design is a form of electronic publishing. This chapter will focus mostly on the principles of design and construction, rather than the specifics associated with the various software packages for coding web pages. However, Macromedia Dreamweaver is arguably the most popular software program used for web design, and thus will be the basis for the few technical sections in this chapter .

Content is the most important aspect of a website, but unlike print media, website content is fluid, which is one of the many characteristics of the web as a publishing medium that makes it so appealing. The ability to constantly edit and update your site can result in an excellent resource and a key way to keep people informed about the latest research without the normal delays associated with print media. Of course, the danger is making information available without the usual peer-review process. However, with appropriate caveats, preliminary findings (not raw data) can be made available to other researchers, thereby promoting early discussion and collaboration.

WHAT IS THE WEB?

The World Wide Web (www) or the 'web' is a system of Internet servers that uses HTTP (Hypertext Transfer Protocol) to transfer specially formatted documents. The documents are formatted in a platform-independent (i.e., works on Windows, Macintosh, UNIX) language called HTML (Hypertext Mark-up Language) that supports links to other documents, as well as graphics, audio, and video files. The web is only a small part of the Internet. The Internet is the vast collection of inter-connected networks that all use the TCP/IP (Transmission Control Protocol/Internet Protocol) protocols. It evolved from the ARPANET (Advanced Research Projects Agency Network) created by the U.S. Department of Defense during the late 1960s and early 1970s.

WHY HAVE A WEBSITE?

The web is an excellent way of communicating your science to the widest possible audience in the shortest timeframe possible, with the ability to edit and update the material an infinite number of times. Publishing your research on the web provides many collaboration opportunities.

WHAT SHOULD YOU HAVE ON YOUR WEBSITE?

A website consists of a homepage (the first page a visitor will see) with links to a variety of sub-pages within your website. These pages may contain a variety of visual and audible elements, including text, graphics, video, and audio files. After content, a clear and consistent navigation system (menu) is the next most important element of a website. If your visitors cannot easily navigate your site, you will not be able to effectively communicate the content.

For scientists wishing to communicate their research findings to a broader audience, their website should at least contain the following elements:

- affiliation (this may be obvious if your pages are nested within your institution's website framework);
- research interests;
- current projects (be wary of this one—keep it up-to-date);
- key research findings;
- list of publications (with details for obtaining electronic or printed copies);
- PDFs of your non-copyrighted publications, such as project reports, posters, newsletters, PowerPoint presentations;
- curriculum vitae; and
- contact details (this is a must and should be an *obvious* part of your navigation system—you are trying to promote collaboration and communication of your science, and interested people must be able to contact you).

The ability for regular editing of websites, compared with a final printed document, can result in a website with out-of-date information or broken links—when you have linked to another website which has since changed or deleted pages. This is something that you also need to consider when planning the content of your pages. How much time do you have to devote to maintaining and updating your pages? If the answer is "not much", then you should consider avoiding content that is subject to change, like *Upcoming events*.

Posting copyright material on your website

According to the copyright agreement that you sign with most journals, you cannot post PDFs of your published papers on the web. There are some exceptions—please refer to SHERPA, the publisher copyright policies and self-archiving website *http://www.sherpa.ac.uk/romeo.php*, where you can find a summary of the permissions that are normally given as part of each publisher's copyright transfer agreements. An alternative is to post details of your papers with abstracts and provide your contact details so interested visitors can request electronic or hardcopy reprints.

LAYOUT AND DESIGN

Fancy bells and whistles will not make up for poor design and content—in fact, they will probably decrease the effectiveness of the page to communicate your science.

Once you have decided on the content for your pages, it is a good idea to draw a flow chart showing how the pages will link together. Put your homepage at the top or middle of the page (whatever works best for you conceptually) and then lay out the content pages and draw the links between the various pages. Remember that your navigation system must allow the user to *get to any page from any page*. At this stage you must also consider whether your pages will be integrated into your institution's existing design layout, or whether you or your group will have a completely independent design.

Often the best way to design a web page is to mock-up the layout in a graphics program such as Adobe Illustrator. Using this technique, you can quickly try several layouts and navigation systems without needing to mess with HTML coding and all its idiosyncrasies. During the design process, be conscious of producing a layout that will facilitate the following web design rules.

Maintain a consistent navigation system (menu) across all pages

As an absolute minimum, always provide a consistent 'one click' link back to your homepage.

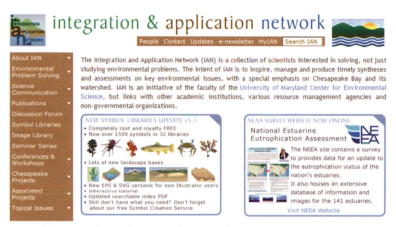

IAN website showing a clean, consistent navigation system common to all pages.

Make hyperlinks obvious

Do not mislead your audience by having elements that look like links unless they are (i.e., underlined text) and make sure all items that are links, are obvious. One of the more annoying elements to the web is navigational ambiguity. One example is 'mystery meat navigation', a technique which uses unlabeled graphics as the menu system.

Use the Open in New Window and pop-up window features wisely

Typically a good rule of thumb is that links to external pages should open in a new window, so visitors do not lose access to your pages. Links to local pages should open in the current window. Exceptions are pages that for some reason do not contain the same navigation (menu) system as the rest of your website. Another option is the use of a pop-up window. Pop-up windows must be user-triggered, and not spawn automatically. Appropriate use for a user-initiated pop-up may be for an online form submission or to display a larger image. These windows are usually fixed in size and do not have many of the standard control bars, thereby making it obvious that they are designed to be closed when the user is finished, returning them to the main website window.

Design your pages to look the same on all browsers

Your pages should be viewable on all *browsers* (i.e., Firefox, Internet Explorer, Netscape, Opera, and Safari), *platforms* (i.e., Windows, Macintosh, Linux, and Sun Solaris) and *screen resolutions* (i.e., 640 x 480, 800 x 600, 1,024 x 768, 1,280 x 1,024, 1,152 x 864, 1,600 x 1,200, 1,680 x 1,050, and 1,920 x 1,200). This is probably the most challenging and most ignored aspect of web page design. All these combinations can be quite a daunting obstacle to universal page design. Fortunately, this is becoming easier with the introduction of xhtml (Extensible html) and css (Cascading Style Sheets). Currently most web pages use a mix of techniques to achieve 'similar' results across all these different barriers. This should become a problem of the past, but at the moment you will have to learn many of the hacked techniques to achieve your design objectives.

IAN website showing some of the problems that can occur with a fixed-width design. This 640 x 480 resolution now accounts for less than 1% of users and so is usually ignored by web developers. This page is correctly displayed on an 800 x 600 resolution screen, which is the minimum resolution still in common use.

Avoid the use of frames

Websites that utilize frames have many disadvantages and very few advantages. Typically, they are used to avoid having to make changes to all your web pages when you need to modify your menu system. They also allow static positioning of title and menu bars, which is a convenient effect, as it keeps the menu available no matter how far the user has scrolled down your page. There are now better ways to achieve these effects, but they do require more complicated coding. Regardless, frames should still be avoided because of the following:

- they are difficult for search engines to navigate (resulting in poor page rankings for your sub-pages);
- if a search engine does index your subpages, the link that a user will follow will bring them to the sub-page without the surrounding frames, leaving them with no navigation system and hence no ability to explore your site any further;
- when a user bookmarks a page on your website, they will only get a link to your homepage—a very frustrating scenario. The same occurs if someone tries to copy the URL (Uniform Resource Locator—the 'www...') to send it to a friend (reducing personal referrals to your site);
- printing framed pages often results in the containing frames being ignored, or infomation beyond the screen height being omitted; and
- frames hinder the use of your site by those who require text-to-speech software to read your site.

Keep colors to a minimum (two or three harmonious colors)

Overuse of color gives a website that 'home-made' feel, which may be a selling point for apple pie, but not for websites. A maximum of three colors is a good rule of thumb.

Use dark text on light-colored backgrounds

The reverse (light text on dark backgrounds) can be hard to read.

The IAN website modified to highlight the poor choice of a dark background. It also uses too many colors.

Use clean, vivid images

Makes sure your JPEGs do not contain compression artefacts and are not blurry. Please see Chapter 3 for more information on correctly preparing images for the web.

Double-check links and spelling

Nothing reduces your credibility more than broken links, poor grammar, and spelling mistakes. Macromedia Dreamweaver and many other HTML editors have built-in spell checking. Use them!

Do not have links to pages that are under construction

Do not post the link until it is ready. It is not good advertising for an upcoming feature—rather, it shows that your website is lacking sufficient content. If you are referring to your whole website, then that is stating the obvious—by their very nature websites are constantly a work in progress. There is no need to state this.

Do not get too fancy

Clean and simple designs will be easier to navigate and will be less prone to the current 'fashions' in web design. Avoid things like splash pages (a page with a pretty banner that has to be clicked on to get to the real homepage), spinning logos, animations (unless they contain meaningful information), 3D graphics, music, and cheap or ugly clip art. For more examples of what to avoid in website design, see *http://www.webpagesthatsuck.com/*.

A couple of examples of images that were common in the early days of the web. There is no excuse for using graphics like these anymore, but unfortunately, they are still present in too many websites.

Do not make long web pages

Excessive scrolling turns visitors away. Try to limit the length of a page to no more than two or three page heights (remember this will vary with different screen resolutions). You can also use internal page links (named anchors—refer to technical section below) to allow the user to skip long sections to reach what they are interested in.

Do not use graphic versions of text to replace headings

Read the section on search engines for the reasons why this is such a bad idea, even though it may look pretty.

WEBSITE ORGANIZATION

There are a few simple rules that will greatly simplify the organization of your website and prevent considerable future editing and re-coding of your pages.

Organize your pages into a folder system as you would the files on your computer

Your website will have a root directory that may be something like: http://www.yourcompany.com/yourname. In this root directory you must have your homepage and it must be called index.html (note: this file must have the full .html extension). This is the file that web browsers will open when someone types in http://www.yourcompany.com/yourname. Also in this root directory you should put your basic first level pages such as contact.htm (note: all other files apart from the index.html file on your website do not need the full .html—most sites use the abbreviated .htm extension).

To keep your site organized, it is common to use a folder (named images) at the first level inside your root directory, which contains all the images on your website. Another example may be a pdfs folder or a videoclips folder. You may also include sub-folders to organize major sub-areas of your site. Some examples may be folders for certain members of your group and for various projects. An example structure might look something like this:

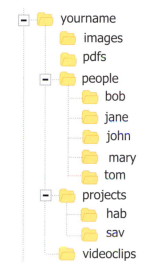

Example file structure for a website. This needs to be exactly the same on your computer and on your web server.

Files within your website can then be linked using relative paths, rather than always having to specify the full URL. This also allows easy transfer from one domain to another without having to rename links.

There are also certain things to avoid in your website folder structure that may be typically used to sort the files on your personal computer. An example would be the use of current projects and finished projects folders. While it might be convenient on your computer and easy to move a project's files over once it is finished, it is not so easy on the web, due to all the links that you would have to modify. To further complicate things, not only will your pages link to the old location, but also, search engines and other websites that have linked to your pages will have outdated links. This leads to the next rule.

Avoid moving your pages around within your file structure

Ideally you would never move anything, but as your website grows and you adopt new design techniques as your skills improve, you may need to make alterations. In this case, it is essential that you use a redirect function from the old page.

If a visitor tries to open a page that no longer exists, they will receive the ubiquitous *404 File Not Found* error page. It is possible to create a custom version of this file so that you can inform the visitor that the contents of the page have been moved and then redirect them to your homepage. This also accounts for links from other sites that may have misspelled the name of your page by directing links to all unknown pages to your homepage. Otherwise, many visitors will assume your whole website is down or no longer available. There are a few different ways to accomplish this and you should contact your information technology department to learn what is appropriate to your server configuration.

All folder and filenames should be lower case and contain no spaces

Use an underscore (_) instead of a space if necessary. Some servers are case-sensitive (UNIX) while others are not (Windows), but it is good practice, as it allows for easy transfer of your site to a different server. Names should also be kept to less than 31 characters (including the .htm, .jpg, etc. extensions), as some Macintosh browsers cannot handle longer filenames.

WEBSITE FORMATTING

Sizing elements on your pages

Try to avoid specifying sizes. Let the reader's browser do the work. If you do specify sizes, *avoid using pixel or point measurements*—use ems or percentages (%) instead. This will allow many of the page's elements to size dynamically— what is known as a fluid layout. The one remaining hurdle to a completely fluid design is the lack of scalability with current graphic formats. A full-scale implementation of the new Scalable Vector Graphics (SVG) will help this problem somewhat, as this format maintains all the scalability functions of the vector format on the web. However, there is still no scalable alternative for photographs, which, as bitmaps, cannot be scaled in size without a concomitant decrease in quality.

Typeface choice

Unlike print media where serif typefaces are designed to create a line that leads the reader's eyes across text, text viewed on-screen is cleaner when a sans serif typeface is used. The other consideration is that you must use a universal typeface that is available on all computers. Common choices in website design are Verdana, Arial, and Trebuchet MS. Most importantly if you are going to set the typeface, be sure to set a few common secondary alternatives, or else the browser's default typeface will be used. Verdana is probably the most ubiquitous typeface on the web, and was designed specifically for readability on-screen at small sizes. However, some critics argue that it is physically bigger than most typefaces at a given pixel size, thereby reducing its aesthetic appearance.

Typeface size

Avoid setting the base typeface size of your pages. Your visitor may need it larger because of poor eyesight or a very high resolution screen. Do not specify typefaces in fixed pixel sizes (e.g., 10 px, 12 px), instead use ems (e.g., 0.8, 1.0, 1.2) or percentages (e.g., 80%, 100%, 120%). This will ensure that your typefaces will appear relative to the base typeface size that the user has set, rather than dictating that to them.

Typeface style

Bold and *italic* should not be the default text in the body of your web pages, but saved for emphasis and other appropriate use, e.g., species names. Underlining should be avoided altogether as it looks like a hyperlink, which will contribute to the navigational confusion of your visitors.

Special characters

There are a number of special characters that can be used on the web, but are not included in the graphical interface of Macromedia Dreamweaver and other HTML editors. Some examples relevant to scientists are included below with the accompanying HTML and CSS code.

NH_4^+　　NH₄⁺

$\delta^{15}N$　　.stylegreek {font-family: Symbol}　　<span class="stylegreek">d</span>¹⁵N

Color

The web used to be designed in so called *web safe colors*, a subset of 216 colors which were common to PC and Macintosh computers. This is typically no longer necessary except for a very small percentage of users on old computers. This restriction meant that photographs never rendered properly. Since the web is a digital medium, colors must be in the RGB color scheme. Initially, the limited abilities of graphics display cards in personal computers meant that the maximum number of colors that could be displayed was 256 (8 bit). Most computers are now capable of displaying 65,000 colors (16 bit) or even 4 billion colors (32 bit).

Colors in HTML are described in hexadecimal codes, rather than RGB or CMYK which are mentioned elsewhere in this manual with respect to print design. Hex codes are made up of six digits and may include 0–9 and A–F, with the letters being used in order to represent the numbers 10–15, each with a single character. Hex codes relate directly to the RGB system, with the first two of the six digits representing the Red, the next two representing the Green, and the last two representing the Blue. 'No color' is represented by 00 and FF represents 100% of that color. Some example codes for colors are included below (note that codes are preceded by the # sign).

#FFFFFF = white

#000000 = black (K)

#FF0000 = red (R)

#00FF00 = green (G)

#0000FF = blue (B)

#FFFF00 = yellow (Y)

#00FFFF = cyan (C)

#FF00FF = magenta (M)

Cascading Style Sheets (CSS)

With the relatively recent introduction of Cascading Style Sheets (CSS), it is now possible to separate information from presentation. The page information is provided using HTML and the presentation (or formatting) is controlled by external CSS code. The main advantage with this approach is the ability to make global changes to formatting by modifying one CSS file rather than modifying the content of every HTML page (this used to be a huge task in large websites which had hundreds of pages). Example of what CSS can control include background page color, typeface size, margin sizes, hyperlink appearance, and table formatting. If you wish to change the appearance of your entire website, you need to change the styles in the CSS file, which controls the appearance of all your HTML pages. If your website is being incorporated into your institution's layout, you can use their CSS file and link your page to it, saving you considerable effort.

Graphics

Quality

Unlike other media, preparing graphics for the web is a balance between quality and download time. The rule of thumb says to keep image file size under 50 KB, preferably under 20 KB (for automatically loading images). In particular avoid large, slow images and other media on your homepage—it does not create a good first impression if your homepage takes too long to load. Larger images can be displayed if users will be clicking specifically to open them (include a warning that they are about to view a large file).

Sizing

Graphics displayed on-screen are different from print in that the size is determined by its dimensions in pixels (px) or points (pt) and has no relation to DPI or physical print size. As an example, an image which is 1,800 x 1,200 px at 300 DPI would be 6″ x 4″ when printed. However, the only element relevant for display on a computer screen is the pixel dimensions, which in this example would be too big, considering the resolution of most screens is 1,024 x 768 px. A more appropriate 'size' might be 300 x 200 px. This would fit well within a web page layout. If printed, this change in pixels would result in a print size of 1″ x 0.667″ at 300 DPI. DPI has no effect on the physical size of what is displayed on a computer screen. In Adobe Photoshop *View* menu, switch between *Print Size* and *Actual Pixels* to get a feel for what is happening here. Also, try this exercise—scan a 35 mm slide at 4,000 DPI and view it in Adobe Photoshop. If you view *Actual Pixels*, the image will be enormous and you will have to scroll around to view the whole image, but if you view *Print Size*, it will appear the size of the original slide. So remember, when you are scanning, that you are scanning a resolution at the current size of the object. When a 35 mm slide scanned at 4,000 DPI is converted to a useful print size (e.g., 8″ x 12″), it will no longer be 4,000 DPI, but closer to 500 DPI. Even though the number of physical pixels has not changed, they are simply taking up more space.

This version is the original 35 mm slide size scanned at 2,200 DPI. It is 2,048 x 3,072 pixels in dimension.

This version has been resized in Adobe Photoshop to 6″ x 4″ with *Resample Image* unchecked. As a result of this change in print size, it is now only 500 DPI. However, it is still physically 2,048 x 3,072 pixels in dimension.

File formats

There are currently two main formats popular on the web—JPEG (JPG) and GIF. The basic rule is that you will get the best quality image at the smallest file size if you use GIF for graphics such as maps, graphs, and diagrams which contain large areas of the same color. GIF format files of simple images are often smaller than the same file would be if stored in JPEG format. The JPEG format is best for photos or graphics which rely heavily on graduated colors.

GIF (Graphics Interchange Format) is a lossless method of compression, meaning that there is no data loss during compression of the image. It uses a simple substitution method of compression. If the algorithm comes across several parts of the image that are the same, e.g., a sequence of digits like this: 1 2 3 4 5, 1 2 3 4 5, 1 2 3 4 5, it makes the letter 'A' stand for the sequence 1 2 3 4 5, so that you could render the same sequence as 'A A A', obviously saving a lot of space. It stores the key to this (A = 1 2 3 4 5) in a hash table, which is attached to the image.

The maximum compression available with a GIF depends on the amount of repetition in an image. A flat color will compress well, while a complex, non-repetitive image like most photos, will not. There are two other key benefits to the GIF format—the ability for transparency so that the graphic effectively has no background and will therefore take on the background color of the web page, and the ability to create simple animations (good for displaying dynamic model outputs). Transparency can easily be created in Adobe Photoshop, and animated GIFs can be created using Adobe Image Ready, a complementary application that comes with Adobe Photoshop.

However, GIFs are limited to a palette of 256 colors or less (8 bit color), making them unsuitable for photographic images and even some diagrams which rely heavily on graduated colors (such as some of the water fills in the IAN symbol libraries). You can also specify the color palette to be used in the image as some graphics may only require four or eight colors to render correctly, thereby reducing the file size even further.

PNG (Portable Network Graphic) image format supports a 24 bit (full color) format similar to GIF, but the continued lack of support for the transparency feature of PNG in Internet Explorer (the dominant web browser) has prevented it from becoming a standard format on the web. The transparency features of the PNG format are far superior to the GIF format and simplify many aspects of web design. Microsoft has recently confirmed that Internet Explorer 7 will support PNG transparency.

JPEG (Joint Photographic Experts Group), by comparison is a lossy compression method. In other words, it throws away parts of an image to save space. The JPEG algorithm divides the image into squares (you can see these squares on badly-compressed JPEGs). Using a Discrete Cosine Transformation, it turns the square of data into a set of curves—some small, some big—that go together to make up the image. This is where the lossy bit comes in—depending on how much you want to compress the image, the algorithm throws away the less significant part of the data (the smaller curves) which adds less to the overall 'shape' of the image. You can determine how much data you want to lose, but you must strike a balance between file size and image quality. Too much loss can create artifacts—unwanted effects such as false color and blockiness.

SVG (Scalable Vector Graphics) is a new standard that provides all the functionality of scalable (vector) images to be viewed through a web browser. Unfortunately the format has been slow to take off with all browsers, still requiring an external plug-in to be able to view them.

SWF (Shockwave Flash) is a format more commonly used for fancy animated advertisements on websites. It can also be used to design your entire website, avoiding most of the complications with cross-platform and cross-browser incompatibilities with HTML. However, web pages in Flash tend to be slow to load, complex to design and as such are best left for art, music, and design websites. However, flash files (like SVG) are vector and can be effectively used to display detailed graphics, thereby allowing the user to zoom in for greater detail with no loss in quality. The current advantage over SVG is that current web stats show that 97% of browsers have support for Flash. For an example, see the large versions of the conceptual diagrams on the LOICZ (Land-Ocean Interactions in the Coastal Zone) webpage on the IAN website http://ian.umces.edu/loicz.htm. Click on one of the standard images to view the large Flash version and then right-click to zoom in. You can then pan around the image using the hand with your mouse.

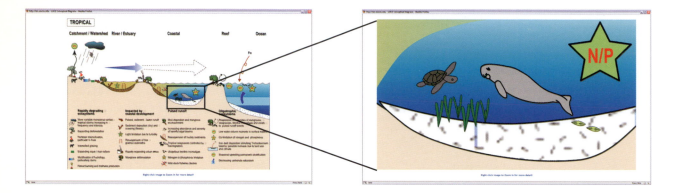

Example of the Land Ocean Interactions in the Coastal Zone webpage on the IAN website making use of Flash to allow for display and zooming of vector images online.

Specifying image sizes

In contrast to most other elements on the web, images should have their height and width parameters specifically set. Re-sizing should be done physically using Adobe Photoshop. Do not use HTML to change the size of images—this results in decreased quality. Also, do not set sizes as percentages of screen size. Although some browsers will correctly render an image without its height and width parameters being set, this sloppiness in adhering to standards is not common among all browsers, so it is good coding technique to follow standards like these. Internet Explorer is renowned for being the most lax when it comes to enforcing standards.

To adjust the pixel dimensions of an image before inserting into your HTML documents, open the image in Adobe Photoshop. Go to *Image → Image Size*. Ensure that *Resample Image* is checked and then adjust the *Pixel Dimensions* to the size you want the image to appear on screen.

Image Size dialog box in Adobe Photoshop.

Saving images for the web

After you have set the appropriate pixel dimensions for the image, use Adobe Photoshop's *File → Save for web* function. From this dialog box you can use various file formats, including GIF, JPG, and PNG. You can also adjust the settings for these formats, e.g., the color palette, dithering, and transparency for GIF, and the image quality for JPGs.

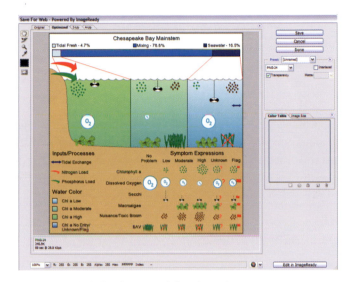

Save for Web dialog box in Adobe Photoshop.

There are many excellent resources on the web for learning Adobe Photoshop and general web graphic optimization techniques. For a list of links, visit the IAN Science Communication Resources → Links page *http://ian.umces.edu/scresources.htm*.

WEBSITE CODING

Now that you have designed your pages (mock-up), created your folder (directory) structure, and prepared your graphics, you can start the coding of your pages. The first step is to set up the site's formatting using CSS. To create a style sheet in Macromedia Dreamweaver, create a new document: *File → New → General → Basic Page → CSS*, which creates a file with a *.css* extension. You then define styles for the various elements on your page using the *CSS Styles* tab. Once you have the basic styles defined you can link it to your HTML pages using the *Attach Style Sheet* command.

Creating your HTML pages

Macromedia Dreamweaver has the ability to place the mock-up image as a background 'tracing image' which you can use to layout your design. Firstly, create a new HTML page using *File → New → General → Basic Page → HTML*. Check the *XHTML compliant* checkbox. Immediately save the document as index.html to the root directory of your file structure. This must be done before inserting any images or linking to other pages, or the relative paths to the files and images will be incorrect. Websites are usually designed and tested locally on the web designer's personal computer, so files should be saved to local directories on your hard drive which replicate the directory structure you will be using on the server.

Websites

Menus

Your menu can be simple text links, links from 'tabs' on an image (using a rollover image technique) or by using drop-down menus where options are revealed on mouseover. Macromedia Dreamweaver includes a built-in system for generating these drop-down menus. There are also several third-party shareware applications which provide more functionality for creating menus. In Dreamweaver you can use *Insert → Image objects → Navigation bar* to create a complete navigational system, or alternatively individual images can have drop menus associated with them using the *Window → Behaviors → Add (+) → Show pop-up menu*. There are also tutorials available on the web to teach you the process of creating drop menus from scratch using layer techniques, which is more time-consuming and complex, but allows much greater artistic freedom.

Tables

Although tables are slowly being replaced by advanced css, they are currently still the easiest way to control content layout on your pages. If table dimensions are set as percentages they will scale with the user's browser window size. Tables are easily constructed and modified in Macromedia Dreamweaver via *Insert → Table*.

Hyperlinks

Hyperlinks are the way your web pages can be linked to other pages on your site and to external websites. To insert a hyperlink in Dreamweaver, use *Insert → Hyperlink*.

- In the *Text* box enter the text that you want to appear on your website as the link to be clicked.
- In the *Link* box, enter the URL of the page into the *Link* box in the *Properties* task pane. If the link is external, be sure to include the http:// followed by the full address. If you are linking to a local page on your site, you can use the *Browse* button to locate the file on your hard drive. If the folder structure is set up to mirror the structure on your webserver, this will create a correct relative reference to the page (i.e., ../contact.htm).
- If you want the page to open in a window other than the current one, you can enter an alternative in the *Target* box. Selecting _blank will open the page in a new window.
- In the *Title* box you can enter a description of the link that will be displayed as a 'tooltip' when the user mouses over the link. The code below shows the <a> (or 'area') tag and the href tag used to define hyperlinks. The title tag defines the 'tooltip' text and the target tag defines what window in which the page will open. If the target tag is omitted, the link will simply open in the current page.

    <a href="http://www.someothersite.com/mainpage.htm" title="Open main page at someothersite.com in a new window" target="_blank">hyperlink</a>

The other elements of the dialog box can usually be ignored. To make a hyperlink from an image embedded on your page, select the image and enter the link, target, and alt tags in the *Properties* frame at the bottom of the Macromedia Dreamwaver page layout. You may also want to set the *Border* option to 'o' to turn off the default border which is applied to images which are hyperlinks. However, remember that this is a visual cue to indicate that the image is a link (see the section on linking and mystery meat navigation).

Email link

Insert a link so your visitors can have their default email program open up a new message with your email address automatically in the *To* box. In Macromedia Dreamweaver, use the *Insert → Email link*. The code for email links uses the same <a href> tag, but the address uses mailto: rather than http:// to tell the browser to open the email client instead of a web page.

    <a href="mailto:me@mycompany.com">Email Me</a>

Images

Images can be inserted into your pages in Dreamweaver using *Insert → Image* command. This defines the image source, the alt tag and the width and height attributes of the image. The alt or 'Alternate' tag displays specified text if the image cannot be displayed for some reason (i.e., slow internet or the user simply has images turned off—not as uncommon as you might expect). They are also used by text-to-speech software engines for the delivery of web content to the visually impaired. They are also very important so that search engines can 'see' what your image shows.

<img src="image.jpg" alt="Image" width="224" height="213" />

Image maps

Image maps are images inserted into a web page which contain 'hotspot' links. Image maps can also be used effectively on maps of a study region with different regions of the map taking the viewer to a page on that region. Image maps can be created in Adobe ImageReady, but Macromedia Dreamweaver also does an excellent job and this keeps the workflow simpler. To create an image map in Macromedia Dreamweaver, simply insert an image into your web page (*Insert → Image*) and then use the *Image Map* tools to define the regions of the image that will function as a hyperlink.

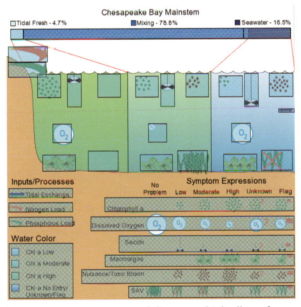

Image map on a conceptual diagram which allows for informative mouseover tooltips and navigation from the diagram. Note that the hashed blue shading does not appear in a web browser. It is shown in Macromedia Dreamweaver to define the area boundaries of the hotspot on the image map.

Video clips and other media

Inserting these elements requires specific code to handle how these play in the user's browser (i.e., whether they start playing automatically, whether they loop on completion, etc.). Use the *Insert → Media* function to select the type of media and then adjust the display settings. Care should be taken not to embed very large files directly into your web page—they should be accessed via a link (with an appropriate warning for low bandwith users) which opens the file in a pop-up window.

Jump menus

Jump menus are the navigational tool common on many websites that provide a drop-down list from which to choose a link. These can be created in Dreamweaver via *Insert → Form → Jump menu*. If you read through the rest of this section, the fields should be self-explanatory.

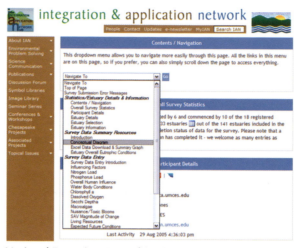

National Estuarine Eutrophication Assessment website utilizing jump menu navigation.

Named anchors

A named anchor can be inserted at a specified point in a page layout using the *Insert → Named anchor* function. You can then link directly to this location by prefixing the name with a pound sign (#) before the name. This technique is useful to navigate through very long pages (although, maybe you should avoid long pages in the first place!).

<a href="#anchorname">Link to Named Anchor</a>

Rollover images

A rollover image is one that changes to something else when the user places their mouse over the image. It is a technique that can be used poorly for 'mystery meat navigation', but can also be quite effective in making a graphic link 'highlighted' to let the user know it is a link to somewhere. In Macromedia Dreamweaver, use *Insert → Image Objects → Rollover Image*. You will then fill in *Image Name* (a reference name for the rollover), browse for the original image (the image appears on the page by default), browse for the rollover image (the image that is displayed when the user mouses over the image), the alternate text (alt tag), and the link (somepage.htm). You should leave the *Preload rollover image* checked as this will ensure that the rollover image is available immediately when the user goes to the link.

Macromedia Dreamweaver's *Insert Rollover Image* dialog box.

User-triggered pop-ups

To create an effective pop-up window, it must be user-triggered, it must not take up the entire screen space, and it should not contain all navigational elements (this clues the user to the fact that it is a window to be closed when they have finished viewing or filling out the form, etc.), but you should retain the ability to resize and scroll the window. Start by creating a hyperlink with the link to # which is known as a null link. You can then use the *Window → Behaviors → Add (+) → Open browser window* to create the pop-up layout. There are a number of parameters to set with respect to the navigational features—use the HELP button for definitions.

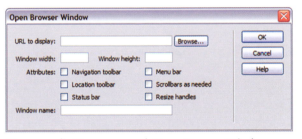

Macromedia Dreamweaver's *Open Browser Window* dialog box for making pop-up windows.

Forms

Forms can be very useful additions to web pages and effective replacements for email links. Forms can be used to ensure the user provides you with specific information and can also give the user pre-defined options. Form results can be emailed to a specified email address and/or can be sent to a database for subsequent retrieval and analysis. The use of forms typically requires a scripting language such as PHP (Hypertext Preprocessor) which is beyond the scope of this manual. However, there are many excellent tutorials on the web for writing PHP form scripts.

MAKING YOUR SITE LIVE

There are a few different ways to actually get your site live, depending on how your server is configured. If your computer is on the same network as the web server you may simply be able to map the root public directory as a network drive on your computer and transfer the files using a normal file manager like Windows Explorer. FTP (File Transfer Protocol) is probably the most common option, although increasingly this is being disabled due to security breaches. If available to you, this system of transferring your files is very simple and quick, but does require specialist software (there are several freeware FTP clients available, e.g., http://www.smartftp.com). Macromedia Dreamweaver also includes a built-in FTP client which automatically uploads files as they are updated locally. Some hosts provide a web interface for uploading your files. This is adequate for small sites, but becomes too slow to manage a large website with many pages. Your entire website with its accompanying directory structure needs to be uploaded to the root directory of the public accessible directory on your web server.

Websites

TIPS AND TRICKS

Stealing from other pages

One of the best ways to learn how to code web pages is by stealing. In the scientific world this may be seen as plagiarism, but on the web this is a normal technique and is encouraged as 'knowledge sharing'. The simplest way is to visit a web page you like and use your browser's *View Source* command. The effectiveness of this technique has been reduced somewhat due to the dynamic nature of many modern web pages, however it is still useful for learning basic techniques. There are also many websites which maintain large databases of sections of free code for accomplishing various tasks. Also, do not forget online discussion forums—chances are that if you are trying to accomplish something, someone else has probably already asked how to do it and received help from the online web design community.

Making your pages cross browser/platform compatible

There are several tools that you can use to test the compatibility of your website.

- The latest version of Macromedia Dreamweaver underlines suspect code in red. If you mouse over the underlined text, it gives you a list of the browser versions that may not be able to properly interpret the code.
- Use the *File → Check page* features of Dreamweaver to test for a variety of coding and accessibility problems.
- There are free validation tools on the web that will scan your pages for errors and give you a report. *http://validator.w3.org* will check your pages to see if they are valid XHTML. *http://jigsaw.w3.org/css-validator* will check your pages to see if they are valid CSS.
- Turn on script debugging. In Internet Explorer, go to *Tools → Internet options → Advanced → Browsing* and uncheck *Disable Script Debugging* and check *Display a notification about every script error*. This will check for errors in your Javascript code that may be ignored by the browser you are using, but may cause your site to completely fail if a visitor is using a more standards-compliant browser. As you surf the web with script debugging on, you will notice that errors are rife. Nevertheless, you should strive to eliminate them on your site.
- Do not use Microsoft Word or other non-specific software to create web pages. Many programs now include the option to export a file as HTML, however this is generally a bad idea. Microsoft Word in particular creates voluminous code with large amounts of proprietary code, which reduces cross-browser compatibility.

Finding out how to do something

Your first resource may be the Macromedia Dreamweaver help file for issues relating specifically to implemeting web design in Dreamweaver, but you should also look at web resources—there are many excellent tutorials available. In fact, the best place to learn about web design and coding is on the web, and the best way to find that information is typically through a Google search (*http://www.google.com*). There are literally thousands of websites containing information, tips, and tutorials on web design. Additionally, there are hundreds of discussion forums dedicated to providing free sharing of information and ideas. Searching through these forums will often bring up the answer you need, but if not, you can ask a question and will often get a reply the same day from someone who is willing to either provide the code you need, or point you to another resource that will help. Google also indexes these forums and will often bring up links to the answer you need directly from your web search.

Search engines

Search engines search web pages, and display results to the searcher. By coding your web pages correctly, you can increase your search engine rankings, thereby increasing the number of hits to your website. This might well be one of the most confusing elements of web design. There are, however, a few simple rules to follow.

- <meta> tags were designed for searches, but are now mostly ignored as they can be abused. Do not leave them off your site, but do not expect them to work miracles. A good title tag is still essential and the description tag is also important.

Title: Document title (appears in the browser's title bar and is also the text that appears when someone bookmarks your page)—up to 50 characters.

Description: This is the sentence that will be displayed in search results below the title of your page. It should be limited to 200 characters.

Keywords: Words separated by commas—up to 1,000 characters.

- Hidden content is now being ignored by search engines, as often it is abused.
- Your domain name is important in search engine rankings so www.yourcompany.com will result in much better rankings than www.freewebhost.com/yourcompany.
- Search engines will read the title of your document, as defined in your <title> tags. Make this meaningful.
- Remember that search engines look for words and phrases that are emphasized in some way, especially the <H1> tags, and to a lesser degree, the bold <strong> and italic <em> (emphasis) tags. This is one very good reason not to use graphic versions of text for your headings.
- Be sure to use <alt> tags for your images or the search engines will not be able to 'see' what the content of the image is. Keep this in mind when designing your pages. Since search engines do not index images, they will not index any text that your website presents in image format, such as a logo. <alt> tags can be thought of as images' descriptions.
- Changing the file structure of your site can kill your rankings. Search engines do not update links every day so you might be waiting up to a month to have your site indexed again after a major change, or even for a new page to appear in the search engine results.
- One of the single most important factors in getting high page rankings—resulting in your page coming up first in the list of search results—is the number of links you are getting from other pages. Also, the more popular the page that links to you, the higher your page ranking will be.
- Remember that high page rankings take time—the more people who visit your site, the higher your page ranking will be, which in turn will bring more people. Use the Google Links feature to see who links to your website.
- Google offers a free service whereby you can place a search box on your website that uses the Google search engine to return results from your website only. This can enhance the ability of users to find information on your website. However, this is not a substitute for a well designed navigational system.

Website statistics

Good website statistics are essential for better targeting your pages to your visitors' interests. Information can include:

- total hits (a fairly useless number);
- unique page visits (the best indicator of the number of visitors to your website);
- which pages were visited;
- how long they spent at your site;
- their entry and exit page;
- whether they bookmarked your site;
- their country;
- which plugins were available in their browser;
- their platform, browser version, and screen resolution;
- the link they came from;
- what keywords have been used in the search engines to direct visitors to your site; and
- any error messages they may have received from your server.

9.

A case study of effective science communication

This chapter is intended to give an overview of the Healthy Waterways Partnership in South East Queensland, Australia, which delivers integrated science, monitoring, planning, and implementation programs supported by extensive government, industry, and community involvement, and targeted communication and education initiatives.

WHAT IS THE HEALTHY WATERWAYS PARTNERSHIP?

Background

The coastal regions and waterways of South East Queensland, Australia, including Moreton Bay, represent unique and complex ecosystems that have a high conservation value and support major recreational and commercial fisheries. The agricultural districts of the region also contribute significantly to the local and regional economy and, together with the growing urban areas, are heavily reliant on the availability of good quality water supplies. However, the human footprint of these activities has led to significant changes in catchment (watershed) hydrology and sediment delivery, declining water quality, and loss of aquatic biodiversity. Nutrients (particularly nitrogen), fine sediments, and to a lesser extent, toxicants (e.g., pesticides and heavy metals) have already been identified as causes of significant environmental problems. Predicted population increases in South East Queensland have the potential to further impact on the ecological and economic health of its waterways and catchments, and there are growing community expectations about reversing the trends in decline in water quality and ecosystem health.

The Healthy Waterways Partnership ('the Partnership') has been established to coordinate the various initiatives required to achieve the *Healthy Waterways/Healthy Catchments* vision in the region. The Partnership delivers integrated science, monitoring, planning, and implementation programs supported by extensive government, industry, and community involvement, and targeted communication and education initiatives. The Partnership framework illustrates a unique integrated approach to water quality management whereby scientific research, community participation, and strategy development are done in parallel with each other. A key attribute is the balance between research, management, and monitoring. The balance ensures a process comprising well-informed management

A case study

decisions based on targeted scientific research and an integrated monitoring program. For stakeholders, it is crucial that an ongoing effort delivers scientific findings in formats suitable to this diversity of groups.

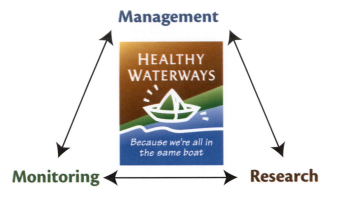

The Healthy Waterways philosophy is to ensure that there is a balance between research, management, and monitoring.

Healthy Waterways/Healthy Catchments vision

The Partnership, through its members, seeks to realize the vision that: *Our waterways and catchments will, by 2020, be healthy ecosystems supporting the livelihoods and lifestyles of people in South East Queensland, and will be managed through collaboration between community, government, and industry.*

The adaptive management approach

With today's increasing emphasis on regional natural resource management, there is a strong need to deliver to resource managers enhanced capacity to make decisions on appropriate management actions. The adaptive management approach is based on the recognition that we often need to act on the basis of an imperfect understanding of the systems within which management action occurs. However, unless there is active research to expand the knowledge base for management as well as appropriate decision support tools for stakeholders and more importantly, the effective communication of this knowledge to stakeholders, the outcomes of the adaptive management process will improve only slowly, if at all. Thus, the Partnership is firmly committed to continually improving the knowledge base and the development of decision support tools to assist stakeholders in the achievement of natural resource management outcomes. The Partnership believes that improved understanding and availability of appropriate decision support tools for management of land and water resources, through effective science communication, result in appropriate prioritization of initiatives targeted towards achieving the vision.

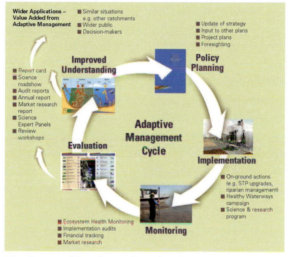

The Partnership adopts the Adaptive Management Framework.

COMMUNICATING SCIENCE: AN ADAPTIVE PROCESS

Effective science communication underpins the four key elements of the Partnership's operating philosophy: stakeholder involvement; adaptive management; continually improving the knowledge base for management; and implementation at the most appropriate level within an integrated regional planning framework.

A significant focus of the work of the Partnership has been the engagement of the scientific community in addressing management needs. Ongoing communication and review, facilitated by the Secretariat, ensures that the science program remains coordinated, relevant, focused, and of high quality. The Partnership has also recognized that information needs to be made available to resource managers, policy-makers, and the wider community in a form that is meaningful to interested parties. For this reason, there is a strong emphasis on the Partnership's work on the accumulation, storage, provision, and communication of management-useful information and decision support tools.

The Science Program of the Partnership is an essential source of information feeding into the Healthy Waterways Campaign, a region-wide public relations and behavioral change initiative of the Partnership, which seeks to deliver the *Healthy Waterways/Healthy Catchments* vision through educating stakeholders to adopt the Partnership objectives.

A case study

The campaign aims to promote the Healthy Waterways brand (the distinctive Healthy Waterways logo with the tagline: *Because we're all in the same boat*). The Healthy Waterways Campaign Program works closely with the other program areas to ensure that communication approaches are relevant to specific audiences.

For the Partnership, effective science communication has been, is, and will continue to be crucial to:

- enhance effectiveness of the Healthy Waterways Campaign;
- ensure communication approaches are relevant to specific audiences;
- facilitate increased community awareness of waterways and catchment issues so as to generate broad support for the *Healthy Waterways/Healthy Catchments* vision;
- increase community action to protect and enhance the health of waterways; and
- increase appreciation of waterways as an important part of lifestyle and culture.

Conceptual diagrams

The Partnership recognizes that a sound understanding of the key processes of the system relative to the priority issues and pollutants identified is necessary for the development of effective management actions. Conceptual diagrams effectively illustrate and synthesize both preliminary and current understanding of the waterways. Using conceptual diagrams to illustrate preliminary understanding of the waterways facilitates the prioritization of issues and consequently, identification of gaps which need to be addressed through research. For example, in 1997, the Stage 2 conceptual diagram depicting four main functional zones (riverine, tidal estuary, seasonal estuary, and marine) was developed prior to the commencement of all Stage 2 tasks. Based on the initial conceptual diagram, a coordinated set of tasks focusing on point sources in Moreton Bay was then developed to address research gaps, including sediment/ nutrient processes, design of an estuarine monitoring program, and benthic flora (1997–1999). At the conclusion of Stage 2 in 1999, the initial conceptual diagram was then revised, depicting the major impacted areas occurring in the river estuaries and the western embayments of Moreton Bay. Key outcomes include zones of sewage impacts, sediment delivery, deposition and resuspension, near-shore seagrass loss, and primary nitrogen limitation.

The revised conceptual diagram highlighted the need for expansion into the freshwater areas of South East Queensland. Stage 3 (1999) focused on catchment sediment and nutrient delivery and processing, design of a freshwater monitoring program, and toxic cyanobacteria (*Lyngbya majuscula*) blooms in Moreton Bay.

Results of South East Queensland's freshwater, estuarine, and Bay scientific tasks were synthesized into conceptual diagrams highlighting key inputs, processes, and biota. A regional perspective of water quality and ecosystem health was developed.

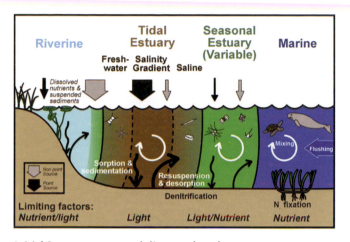

Initial Stage 2 conceptual diagram (1997).

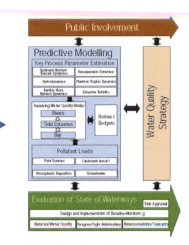

Stage 2 task architecture
(1997–1999).

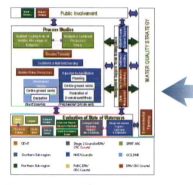

Stage 3 task architecture
(1999–2001).

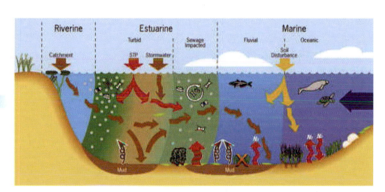

Revised Stage 2 conceptual diagram (1999).

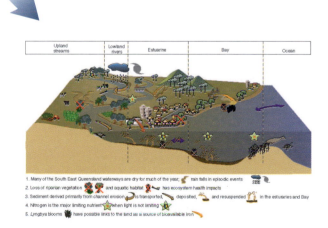

Stage 3 conceptual diagram (2001).

Targeted presentations/workshops

Scientific results and findings and their implications to the development of management actions are regularly disseminated to various stakeholders through public meetings and presentations. In response, stakeholders make commitments to implement management actions after considering the scientific information, community views and issues, results of computer modeling scenarios, as well as the economic, social, and cultural impacts of environmental choices.

Presentations are regularly held by the Partnership to provide updates to stakeholders in order to enhance their current knowledge, facilitate discussion and solicit issues, assist in development and implementation of management actions, justify/request funding (e.g., presentations), seek feedback (e.g., peer review process), and showcase the Partnership activities (e.g., River*symposium*).

In the development of scientific presentations, the Partnership considers the following points:

- type of audience (e.g., scientists/researchers, council technical officers, mayors, general public, or a focused group);
- general purpose/expected outcomes of the presentation (e.g., update of the outcomes of certain tasks, justification for funding or request for future funding, showcase of the Partnership's strengths, etc.);
- time given to do the presentation (e.g., 15-minute presentation vs. one hour); and
- venue and facilities (local council office, university lecture theater, church hall, etc.).

HOW EFFECTIVE SCIENCE COMMUNICATION CAN RESULT IN GOOD MANAGEMENT DECISIONS

Scientific results: Distinct sewage-impacted areas in western and southern Moreton Bay were observed, with additional distinct plumes in Bramble Bay from Pine River/Hayes Inlet and Brisbane River. This was indicated in sewage plume maps generated with stable isotope analysis (δ^{15}N) of marine plant tissue.

Science communication: Results were presented to the various local councils. In particular, the coastal councils of Redcliffe City Council,

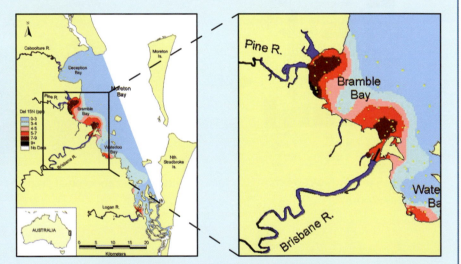

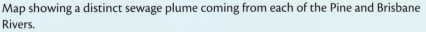

Map showing a distinct sewage plume coming from each of the Pine and Brisbane Rivers.

Pine Rivers Shire Council, and Brisbane City Council recognized the impacts of their individual wastewater discharges and have given their commitment for sewage treatment plant (STP) upgrades to improve Bramble Bay. In the past, Redcliffe City Council and Pine Rivers Shire Council, both of which discharge into Bramble Bay, have always presumed that all the sewage in Bramble Bay was coming from Brisbane River/Luggage Point STP, the biggest sewage treatment plant managed by Brisbane City Council. However, the sewage plume map delineated consistent and separate sewage plumes coming from Brisbane River and Pine River/Hayes Inlet into Bramble Bay. These results further strengthened the effectiveness of a collaborative effort between the different stakeholders in achieving healthy waterways.

Sustainable point source nitrogen loads have been determined for various waterways including Deception, Bramble and Waterloo bays. Best Practice Environmental Management will be achieved at almost all sewage treatment plants and major industrial discharges. This includes an average total nitrogen concentration of 5 mg l^{-1} in treated effluent from the Redcliffe STP by 2005 and 10 mg l^{-1} from Luggage Point STP by December 1999. Further improvements for Luggage Point are proposed by 2005.

Information-based written products (books, reports, newsletters, brochures, report cards, etc.)

On-going delivery of informative, current information suitable for stakeholders and scientists is achieved through the Partnership's written products. Most of these products are distributed as printed copies and posted on the web (*www.healthywaterways.org*). Communication of scientific data is achieved through the use of a diverse range of tools.

Every year, the Partnership also releases the *Ecosystem Health Report Card*, which provides A to F grade ratings for waterways of South East Queensland and Moreton Bay. The Report Card is a culmination of 12 months of monitoring at 120 freshwater and more than 240 estuarine and marine sites throughout South East Queensland. The Report Card is an easy-to-understand snapshot of the Partnership's Ecosystem Health Monitoring Program, which uses rigorous science to identify waterway health using a range of biological, physical, and chemical indicators. By comparing ratings over the years, the Report Card provides an evaluation of the effectiveness of investments in waterway and catchment management undertaken by the community, local and state government agencies, and industry.

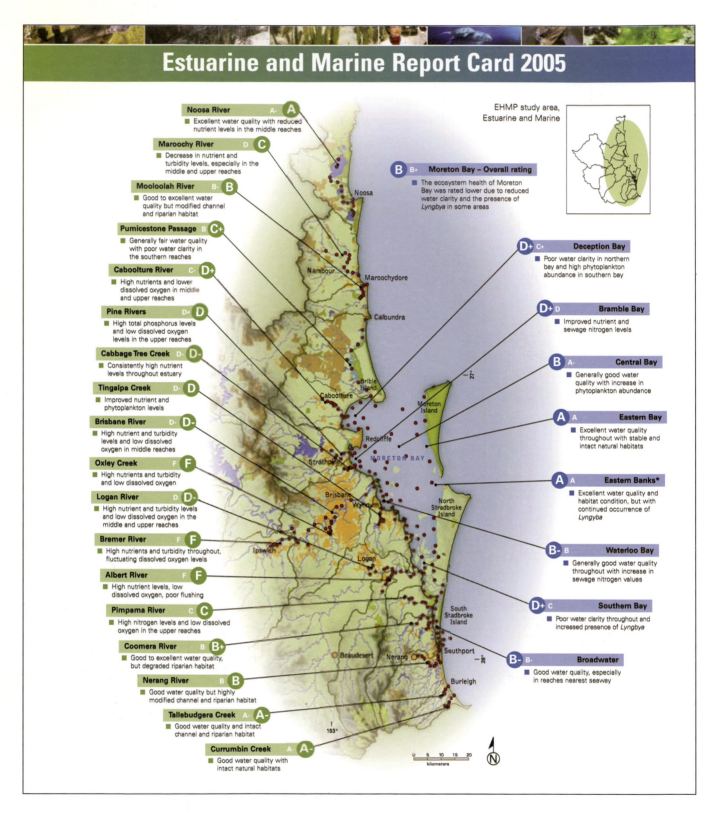

Estuarine and Marine Report Card 2005

Noosa River A- **A**
- Excellent water quality with reduced nutrient levels in the middle reaches

Maroochy River D **C**
- Decrease in nutrient and turbidity levels, especially in the middle and upper reaches

Mooloolah River B- **B**
- Good to excellent water quality but modified channel and riparian habitat

Pumicestone Passage B **C+**
- Generally fair water quality with poor water clarity in the southern reaches

Caboolture River C- **D+**
- High nutrients and lower dissolved oxygen in middle and upper reaches

Pine Rivers D+ **D**
- High total phosphorus levels and low dissolved oxygen levels in the upper reaches

Cabbage Tree Creek D- **D-**
- Consistently high nutrient levels throughout estuary

Tingalpa Creek D- **D**
- Improved nutrient and phytoplankton levels

Brisbane River D- **D-**
- High nutrient and turbidity levels and low dissolved oxygen in middle reaches

Oxley Creek F **F**
- High nutrients and turbidity and low dissolved oxygen

Logan River D **D-**
- High nutrient and turbidity levels and low dissolved oxygen in the middle and upper reaches

Bremer River F **F**
- High nutrients and turbidity throughout, fluctuating dissolved oxygen levels

Albert River F **F**
- High nutrient levels, low dissolved oxygen, poor flushing

Pimpama River C **C**
- High nitrogen levels and low dissolved oxygen in the upper reaches

Coomera River B **B+**
- Good to excellent water quality, but degraded riparian habitat

Nerang River B **B**
- Good water quality but highly modified channel and riparian habitat

Tallebudgera Creek A- **A-**
- Good water quality and intact channel and riparian habitat

Currumbin Creek A **A-**
- Good water quality with intact natural habitats

EHMP study area, Estuarine and Marine

B B+ **Moreton Bay – Overall rating**
- The ecosystem health of Moreton Bay was rated lower due to reduced water clarity and the presence of *Lyngbya* in some areas

D+ C+ **Deception Bay**
- Poor water clarity in northern bay and high phytoplankton abundance in southern bay

D+ D **Bramble Bay**
- Improved nutrient and sewage nitrogen levels

B A- **Central Bay**
- Generally good water quality with increase in phytoplankton abundance

A A **Eastern Bay**
- Excellent water quality throughout with stable and intact natural habitats

A A **Eastern Banks***
- Excellent water quality and habitat condition, but with continued occurrence of *Lyngyba*

B- B **Waterloo Bay**
- Generally good water quality throughout with increase in sewage nitrogen values

D+ C **Southern Bay**
- Poor water clarity throughout and increased presence of *Lyngbya*

B- B- **Broadwater**
- Good water quality, especially in reaches nearest seaway

South East Queensland Estuarine and Marine Ecosystem Health Report Card 2005.

Freshwater Report Card 2005

EHMP study area,
South East Queensland, Australia

Legend
- Catchment border
- Town
- Urban areas
- Protected areas
- State forest
- Monitoring sites
- (A) Excellent
- (B) Good
- (C) Fair
- (D) Poor
- (F) Fail

Waterway name (A) (B)
2004 grade | 2005 grade
*Data from fewer than 5 sites

(B) A- Noosa Catchment
- Streams generally in good condition; lower scores for physical-chemical, ecosystem processes and nutrient cycling indicators in autumn

(C+) C- Maroochy Catchment
- Streams generally in fair condition; higher spring scores than last year for physical and chemical, ecosystem processes and macroinvertebrate indicators

(B-) A- Mooloolah Catchment
- Streams generally in good condition; lower scores than last year for physical-chemical and macroinvertebrate indicators

(C+) C Pumicestone Catchment*
- Streams generally in fair condition; slightly higher spring scores for nutrient cycling and ecosystem processes than last year

(B-) C- Caboolture Catchment
- Streams generally in good condition; improved scores for most indicators in spring and nutrient cycling in autumn

(C) D Pine Catchment
- Streams generally in fair condition; improved spring scores for ecosystem processes, macroinvertebrate and fish indicators

(D-) F Lower Brisbane Catchment
- Streams generally in poor condition; slightly increased scores for nutrient cycling and macroinvertebrate indicators

(F) D Redlands Catchment
- Streams generally in very poor condition; consistently low scores for nutrient cycling, aquatic macroinvertebrates and fish indicators

(B+) A- Nerang Catchment
- Streams in very good condition; inclusion of previously unavailable nutrient cycling data for spring

(C-) B+ Tallebudgera/Currumbin Catchments*
- Streams in fair condition; inclusion of previously unavailable data for the nutrient cycling indicator

Stanley-Kilcoy Catchment (B) (B)
- Streams remain in good condition; no substantial changes

Upper Brisbane Catchment (C) C-
- Streams remain in fair condition; scores for the nutrient cycling indicator consistently amongst the best scores for the Upper Brisbane reporting area

Mid-Brisbane Catchment* (B-) C+
- This site is in fair condition; reduced score for the fish indicator during autumn

Lockyer Catchment (D) D-
- Streams remain in poor condition; lower scores for most indicators in autumn

Bremer Catchment (D-) D-
- Streams remain in poor condition; most indicators scored poorly throughout the year

Logan Catchment (C) D
- Streams generally in poor condition; reduced scores for both nutrient cycling and fish indicators in comparison to last year

Albert Catchment* (B) (B)
- Streams remain in good condition; all indicators scored well in both seasons with the exception of the nutrient cycling indicator

Pimpama/Coomera Catchments* (B) B+
- Streams generally in very good condition; slightly improved scores for ecosystem processes and macroinvertebrate indicators

0 10 20 30
kilometres

153°E

N

South East Queensland Freshwater Ecosystem Health Report Card 2005.

In addition to the report card, the Partnership has synthesized scientific results in a variety of publications, some of which follow.

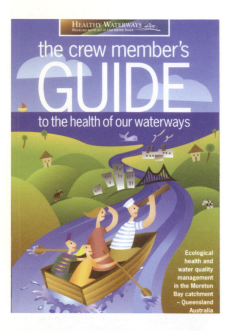

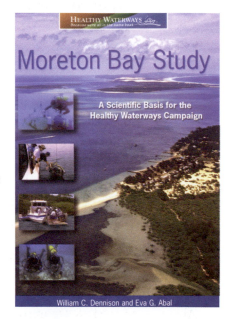

The crew member's guide to the health of our waterways invites the reader to come on board and join the crew to understand and improve our waterways. It introduces the waterways of the Moreton Bay region and its water quality and ecosystem health issues.[1]

The Moreton Bay study summarizes the scientific findings from the Moreton Bay scientific tasks into an issues-based book. Results are illustrated with 181 photographs, 137 maps, 81 diagrams, 125 figures, and 26 tables. Key findings are synthesized onto the regional conceptual diagram.[2]

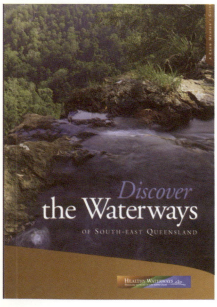

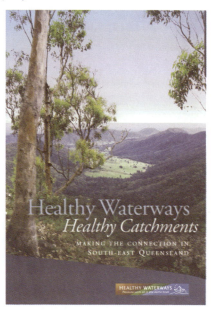

Having introduced the local waterways and discussed regional issues in previous publications, *Discover the waterways of South East Queensland* takes the reader to key sites describing how to get there and what you can see. The publication encourages the reader to go out and see first-hand the condition of South East Queensland waterways. Conceptual diagrams illustrate what can be seen at each of these sites/catchments.[3]

Healthy waterways, healthy catchments: making the connection in South East Queensland provides the context and scientific findings of Stage 3 and serves as a companion book to the *Discover the waterways of South East Queensland* book. The scope of this book reflects the larger 'footprint' into which the *Healthy Waterways Campaign* has evolved, which coincides more closely with the 'ecological footprint' of the South East Queensland region.[4]

CONCLUSIONS

The Partnership has produced information-based outcomes which have led to significant cost savings in water quality management by its stakeholders. This has been achieved by providing a clear focus for management actions that has ownership of governments, industry, and community. As management actions are implemented, the tangible outcomes will be an improvement in water quality for Moreton Bay and its waterways. Scientific researchers were targeted to address issues requiring appropriate management actions, and hence were linked closely to the end users. The outcomes of the Partnership so far have led to significant cost savings in water quality management by government (~$500 million). The management actions proposed were based on a sound understanding of the waterways, a rigorous public consultation/involvement program, and a strategy development incorporating commitments from all levels of the stakeholders. In all of these, the role of effective scientific communication was crucial.

REFERENCES

1. Moreton Bay Catchment Water Quality Management Strategy Team. 1998. *The crew member's guide to the health of our waterways.* Moreton Bay Catchment Water Quality Management Strategy Team, Brisbane, Queensland, Australia.
2. Dennison WC, & Abal EG. 1999. *Moreton Bay Study: A scientific basis for the Healthy Waterways campaign.* South East Queensland Regional Water Quality Management Strategy, Brisbane, Queensland, Australia.
3. South East Queensland Regional Water Quality Management Strategy Team. 2001. *Discover the Waterways of Southeast Queensland* South East Queensland Regional Water Quality Management Strategy, Brisbane, Queensland, Australia.
4. Abal EG, Bunn SE, & Dennison WC (eds). 2005. *Healthy Waterways, Healthy Catchments: Making the connection in South East Queensland, Australia.* Moreton Bay Waterways and Catchments Partnership, Brisbane, Queensland, Australia.

FURTHER INFORMATION

Healthy Waterways   *www.healthywaterways.org*

A case study

Effective communication is 20% what you know, and 80% how you feel about what you know.

Jim Rohn

10.

Index